Sustainability
and the
U.S. EPA

Committee on Incorporating Sustainability in the
U.S. Environmental Protection Agency

Science and Technology for Sustainability Program
Policy and Global Affairs Division

NATIONAL RESEAR

D0891076

THE NATIONAL ACADEMIES PRESS
Washington, D.C.
www.nap.edu

7-17-12
LB
$ 37,00

THE NATIONAL ACADEMIES PRESS 500 Fifth Street, N.W. Washington, D.C. 20001

NOTICE: The project that is the subject of this report was approved by the Governing Board of the National Research Council, whose members are drawn from the councils of the National Academy of Sciences, the National Academy of Engineering, and the Institute of Medicine. The members of the committee responsible for the report were chosen for their special competences and with regard for appropriate balance.

This study was supported by Contract No. EP-C-09-003 between the National Academy of Sciences and the U.S. Environmental Protection Agency. Any opinions, findings, conclusions, or recommendations expressed in this publication are those of the author(s) and do not necessarily reflect the views of the organizations or agencies that provided support for the project.

International Standard Book Number-13: 978-0-309-21252-6
International Standard Book Number-10: 0-309-21252-9

Additional copies of this report are available from the National Academies Press, 500 Fifth Street, N.W., Lockbox 285, Washington, D.C. 20055; (800) 624-6242 or (202) 334-3313 (in the Washington metropolitan area); Internet, *http://www.nap.edu*.

THE NATIONAL ACADEMIES
Advisers to the Nation on Science, Engineering, and Medicine

The **National Academy of Sciences** is a private, nonprofit, self-perpetuating society of distinguished scholars engaged in scientific and engineering research, dedicated to the furtherance of science and technology and to their use for the general welfare. Upon the authority of the charter granted to it by the Congress in 1863, the Academy has a mandate that requires it to advise the federal government on scientific and technical matters. Dr. Ralph J. Cicerone is president of the National Academy of Sciences.

The **National Academy of Engineering** was established in 1964, under the charter of the National Academy of Sciences, as a parallel organization of outstanding engineers. It is autonomous in its administration and in the selection of its members, sharing with the National Academy of Sciences the responsibility for advising the federal government. The National Academy of Engineering also sponsors engineering programs aimed at meeting national needs, encourages education and research, and recognizes the superior achievements of engineers. Dr. Charles M. Vest is president of the National Academy of Engineering.

The **Institute of Medicine** was established in 1970 by the National Academy of Sciences to secure the services of eminent members of appropriate professions in the examination of policy matters pertaining to the health of the public. The Institute acts under the responsibility given to the National Academy of Sciences by its congressional charter to be an adviser to the federal government and, upon its own initiative, to identify issues of medical care, research, and education. Dr. Harvey V. Fineberg is president of the Institute of Medicine.

The **National Research Council** was organized by the National Academy of Sciences in 1916 to associate the broad community of science and technology with the Academy's purposes of furthering knowledge and advising the federal government. Functioning in accordance with general policies determined by the Academy, the Council has become the principal operating agency of both the National Academy of Sciences and the National Academy of Engineering in providing services to the government, the public, and the scientific and engineering communities. The Council is administered jointly by both Academies and the Institute of Medicine. Dr. Ralph J. Cicerone and Dr. Charles M. Vest are chair and vice chair, respectively, of the National Research Council.

www.national-academies.org

COMMITTEE ON INCORPORATING SUSTAINABILITY IN THE U.S. ENVIRONMENTAL PROTECTION AGENCY

Bernard D. Goldstein, M.D., *(Chair)*, University of Pittsburgh, Pennsylvania
Leslie Carothers, L.L.B., L.L.M., Environmental Law Institute, Washington, D.C.
J. Clarence Davies, Ph.D., Resources for the Future, Washington, D.C.
John Dernbach, J.D., Widener University School of Law, Harrisburg, Pennsylvania
Paul Gilman, Ph.D., Covanta Energy Corporation, Fairfield, New Jersey
Neil Hawkins, Sc.D., The Dow Chemical Company, Midland, Michigan
Michael Kavanaugh, Ph.D., P.E., Geosyntec Consultants, Oakland, California
Stephen Polasky, Ph.D., University of Minnesota, St. Paul, Minnesota
Kenneth G. Ruffing, Ph.D., Independent Consultant, Paris, France
Armistead G. Russell, Ph.D., Georgia Institute of Technology, Atlanta, Georgia
Susanna H. Sutherland, M.S., City of Knoxville, Tennessee
Lauren Zeise, Ph.D., California Environmental Protection Agency, Oakland, California

Science and Technology for Sustainability Program Staff

Marina Moses, DrPH, Director
Dominic Brose, Associate Program Officer
Jennifer Saunders, Program Officer
Dylan Richmond, Research Assistant
Patricia Koshel, Senior Program Officer
Emi Kameyama, Program Associate
Ruth Crossgrove, Senior Editor
Mirsada Karalic-Loncarevic, Manager, Technical Information Center

Preface

Recognizing the importance of sustainability to its work, the U.S. Environmental Protection Agency (EPA) has been examining applications in a variety of areas in order to better incorporate sustainability into decision making at the agency. The agency has also undertaken several sustainability initiatives, and can claim success in developing processes leading to sustainability. However, to further strengthen the analytic and scientific basis for sustainability as it applies to human health and environmental protection, EPA asked the National Research Council (NRC) to convene a committee under the Science and Technology for Sustainability Program (STS) to provide an operational framework for integrating sustainability as one of the key drivers within the regulatory responsibilities of EPA. Specifically, in addition to being tasked with developing an operational framework for sustainability for EPA, the committee was asked to address how the existing framework rooted in the risk assessment/risk management paradigm can be integrated under the sustainability framework; identify the scientific and analytical tools needed to support the framework; and identify the expertise needed to support the framework.

In this report, Chapter 2 first provides a brief history of sustainability, Chapter 3 presents the proposed sustainability framework for EPA, and Chapter 4 discusses the processes and tools to support the proposed framework. In Chapters 5 and 6, the committee provides guidance about how the EPA decision-making process rooted in the risk assessment/risk management paradigm can be integrated into this new sustainability framework and includes a discussion of cultural "change management" at the agency. Finally, Chapter 7 closes by examining the relevance and utility of sustainability considerations in EPA's accomplishment of its mission.

This report has been reviewed in draft form by individuals chosen for their diverse perspectives and technical expertise, in accordance with procedures approved by the National Academies' Report Review Committee. The purpose of this independent review is to provide candid and critical comments that will assist the institution in making its published report as sound as possible and to ensure that the report meets institutional standards for objectivity, evidence, and responsiveness to the study charge. The review comments and draft manuscript remain confidential to protect the integrity of the process.

We wish to thank the following individuals for their review of this report: Michael Callahan, MDB, Inc.; Linda Fisher, E. I. du Pont de Nemours & Company; H. Christopher Frey, North Carolina State University; Howard Frumkin, University of Washington; Gerald Galloway, University of Maryland; F. Henry Habicht, SAIL Capital Partners; Ciannat Howett, Emory University; Pamela Matson, Stanford University; Kathleen McGinty, Weston Solutions Inc.; Hendrik Wolff, University of Washington; Terry Yosie, World Environment Center; and Rae Zimmerman, New York University.

Although the reviewers listed above have provided many constructive comments and suggestions, they were not asked to endorse the conclusions or recommendations, nor did they see the final draft of the report before its release. The review of this report was overseen by Robert Frosch, Harvard University. Appointed by the National Academies, he was responsible for making certain that an independent examination of this report was carried out in accordance with institutional procedures and that all review comments were carefully considered. Responsibility for the final content of this report rests entirely with the authoring committee and the institution.

The committee gratefully acknowledges Paul Anastas, Alan Hecht, Jim Jones, John Frece, Ira Leighton, Mathy Stanislaus, Randy Hill, and Jared Blumenfeld, of the U.S. Environmental Protection Agency; Charles Powers, Vanderbilt University; Ellen Gilinsky, Virginia Department of Environmental Quality; Justin Johnson, Vermont Department of Environmental Conservation; Ann Klee, General Electric Company; Linda Fisher, DuPont; Deborah Swackhamer, University of Minnesota; and E. Donald Elliott, Yale Law School for making presentations to the committee.

The committee is also grateful for the assistance of NRC staff in preparing this report. Staff members who contributed to this effort are Marina Moses, director of the Science and Technology for Sustainability Program; Dominic Brose, associate program officer; Jennifer Saunders; program officer; Dylan Richmond, research assistant; Patricia Koshel, senior program officer; Emi Kameyama, program associate; Ruth Crossgrove, senior editor; and Mirsada Karalic-Loncarevic, manager of the Technical Information Center.

We thank especially the members of the committee for their tireless efforts throughout the development of this report.

Bernard D. Goldstein, *Chair*
Committee on Incorporating Sustainability
in the U.S. Environmental Protection Agency

Contents

SUMMARY   1

1   INTRODUCTION   7
    EPA's Mission and Role, 8
    Committee's Task, 11
    Committee's Approach to the Task, 11
    Structure of the Report, 13
    References, 14

2   HISTORY OF SUSTAINABILITY   15
    Conservation in the United States, 16
    Environmental Protection in the United States, 17
    Sustainable Development, 19
    Findings and Recommendations, 29
    References, 29

3   A SUSTAINABILITY FRAMEWORK FOR EPA   35
    Introduction, 35
    The Sustainability Framework, 36
    The Sustainability Framework: Level 1 Components, 40
    Findings and Recommendations, 49
    References, 50

4   SUSTAINABILITY ASSESSMENT AND MANAGEMENT:
    PROCESS, TOOLS, AND INDICATORS   53
    Elements of Sustainability Assessment and Management, 53
    Findings and Recommendations, 72
    References, 74

5   HOW RISK ASSESSMENT AND RISK MANAGEMENT
    RELATE TO THE SUSTAINABILITY FRAMEWORK   79
    Inference Guidelines and Operational Procedures, 81
    Limitations of the Risk Assessment and Risk Management Paradigm, 82
    Evolution of the Risk Assessment and Risk Management Paradigm, 82
    The Interface Between the Risk Assessment and Risk Management
        Paradigm and Sustainability, 86
    Findings and Recommendations, 89
    References, 90

6   CHANGING THE CULTURE IN EPA   93
    Effecting Cultural Change in the Agency, 93
    Research and Development, 96
    Findings and Recommendations, 106
    References, 108

7   BENEFITS OF A SUSTAINABILITY APPROACH AT EPA   111
    Daunting Challenges, 112
    Probable Benefits of a More Robust Approach to Sustainability, 114
    Conclusion: The Journey Ahead, 117
    Findings and Recommendations, 122
    References, 123

APPENDIXES

A   THE COMMITTEE ON INCORPORATING SUSTAINABILITY
    IN THE U.S. ENVIRONMENTAL PROTECTION AGENCY   127
B   STATEMENT OF TASK   133
C   GLOSSARY   135
    References, 137
D   SUSTAINABILITY IN THE OECD   139
    References, 140
E   SUSTAINABILITY INDICATORS   143
    References, 149

TABLES, BOXES, AND FIGURES

Tables

5-1   Differences Between Features of Risk Assessment and Risk Management and Sustainable Development, 88

E-1   Policy-Based Sustainable Development Indicators, 145
E-2   Outcome-Oriented Sustainable-Development Indicators, 146

Boxes

2-1   International Sustainable Development Conferences, 22
2-2   2009 Monitoring Report of EU Sustainable-Development Strategy, 24

3-1   Examples of Management System Frameworks for Sustainability, 39
3-2   Selected International and National Sustainability Principles, 42
3-3   Goal, Indicator, and Metric, 48

4-1   Biofuels, 54
4-2   Scenarios for Global Biodiversity, 65
4-3   Indicator Attributes, 70

6-1   Everglades Restoration: The Comprehensive Everglades Restoration Project, 94
6-2   Redevelopment of Boston's Fairmount Rail Corridor: Addressing Environmental Justice Issues Through Multi-Agency and Community Collaboration, 100
6-3   Approving New York City's Water Supply Protection System, 101
6-4   An Example of EPA's Role in Facilitating State Activities that Achieve Environmental Goals: Improving Air Quality Through Land-Use Planning, 102
6-5   Growing Collaboration on Redesigning Roofs, 103
6-6   Presidential Green Chemistry Awards, 105

7-1   Green Infrastructure: Sustainable Water Quality Solutions for Cities with Combined Sewer and Storm-Sewer Overflows, 115
7-2   Using EPA Technical Assistance to Aid Advances in Stormwater Best Practices, 116
7-3   Energy Savings Permit Cleanup of Polychlorinated Biphenyl (PCB) Contamination in New York City Schools Without Layoff of Teachers, 117
7-4   Sustainable Solutions to Air Pollution Associated with Public Transit Bus Depots in Northern Manhattan: An Environmental Justice Issue and Community Response, 118

7-5   The Sustainability Remediation Forum: A Private Sector Effort to
       Incorporate Sustainable Practices into Remediation Efforts, 119
7-6   IBM: Early Mover on Corporate Responsibility and Sustainability, 120
7-7   Climate Change Mitigation and Sustainability, 124

Figures

S-1   A framework for EPA sustainability decisions, 4

3-1   A framework for EPA sustainability decisions, 37
3-2   A framework for EPA sustainability decisions (Level 1), 38
3-3   Reorganization themes of the Office of Research and Development, 46

4-1   A framework for EPA sustainability decisions (Level 2), 54

5-1   Elements of risk assessment and risk management in the Red Book, 80
5-2   Framework for risk-based decision making, 85
5-3   Correspondence between sustainability assessment and management
       elements and risk assessment and risk management (RA/RM)
       framework, 89

Summary

Sustainability is based on a simple and long-recognized factual premise: Everything that humans require for their survival and well-being depends, directly or indirectly, on the natural environment.[1] The environment provides the air we breathe, the water we drink, and the food we eat. It defines in fundamental ways the communities in which we live and is the source for renewable and non-renewable resources on which civilization depends. Our health and well-being, our economy, and our security all require a high quality environment.

The U.S. Environmental Protection Agency (EPA) has been working to create programs and examining applications in a variety of areas to better incorporate sustainability into decision making at the agency. To further strengthen the analytic and scientific basis for sustainability as it applies to human health and environmental protection, EPA asked the National Research Council (NRC) to convene a committee under the Science and Technology for Sustainability Program to provide an operational framework for integrating sustainability as one of the key drivers within the regulatory responsibilities of EPA. Specifically, the committee was tasked to answer four key questions:

- What should be the operational framework for sustainability for EPA?
- How can the EPA decision-making process rooted in the risk assessment/ risk management (RA/RM) paradigm be integrated into this new sustainability framework?

---

[1] Marsh, G.P. 1864. Man and Nature; or, Physical Geography as Modified by Human Action. Cambridge, MA: Belknap Press of Harvard University Press.

1

- What scientific and analytical tools are needed to support the framework?
- What expertise is needed to support the framework?

The NRC has looked in depth at the use of the RA/RM framework as a decision-making tool at EPA.[2] This study was to build on that in answering these four key questions. EPA has undertaken several sustainability initiatives and can claim success in developing processes leading toward a more sustainable future. EPA has established various programs incorporating sustainability at the program office and regional level and has adopted a sustainability research plan and highlighted sustainability in its strategic plan for 2011–2015; however, the agency recognizes that to obtain the benefits of using sustainability as a process and as a goal, it needs to further improve and institutionalize sustainability. Paul Anastas, the assistant administrator for research and development at EPA, stated, "Sustainability is our true north. The work that we do—the research, the assessments, the policy development—is part of ensuring that we have a sustainable society; a sustainable civilization."[3]

The growing identification of sustainability as both a process and a goal to ensure long-term human well-being is based on four converging drivers. The first is the recognition that current approaches aimed at decreasing existing risks, however successful, are not capable of avoiding the complex problems in the United States and globally that threaten the planet's critical natural resources and put current and future human generations at risk, including population growth, the widening gaps between the rich and the poor, depletion of finite natural resources, biodiversity loss, climate change, and disruption of nutrient cycles. Second, sophisticated tools are increasingly available to address cross-cutting, complex, and challenging issues that go beyond the current approach, which is, risk management of major threats. Third, sustainability is being used by international organizations as a common approach to address the three sustainability pillars (social, environmental, and economic issues) in agreements in which the United States is an active participant. Finally, the potential economic value of sustainability to the United States is recognized to not merely decrease environmental risks but also to optimize the social and economic benefits of environmental protection.

To accomplish its task of answering the four key questions posed by EPA, the committee held meetings in December 2010 and February 2011 and conducted weekly conference calls to discuss the report draft. The February meeting was a weeklong intensive session, which included extensive review and discussion of relevant literature, deliberation, and drafting of the report. In addition, data-gathering sessions that were open to the public were held during both meetings, where the committee heard from EPA officials, state-level environmental agencies, industry,

---

[2] NRC (National Research Council). 1983. Risk Assessment in the Federal Government: Managing the Process. Washington, DC: National Academy Press.

[3] Anastas, P. 2010. Testimony before the U.S. House of Representatives, Committee on Science and Technology March 10, 2010 [online]. Available: http://www.epa.gov/ocirpage/hearings/testimony/111_2009_2010/2010_0310_pa.pdf [accessed Apr. 19, 2011].

universities, and nongovernmental organizations that provided a range of perspectives on sustainable development and environmental stewardship. The committee addressed its task by providing guidance to EPA on the processes necessary to incorporate sustainability into the agency's work. The committee did not provide guidance on the specific direction the agency should take to accomplish this task.

The committee reviewed a large body of written material on sustainability in the United States, as well as internationally, and reviewed many other documents related to EPA's structure and function. The committee also did not devote significant time to defining sustainability but used the definition from Executive Order 13514, where it is defined as

Sustainability: "to create and maintain conditions, under which humans and nature can exist in productive harmony, that permit fulfilling the social, economic, and other requirements of present and future generations" (NEPA[1969]; E. O.13514[2009][4]).

The committee has not examined whether or to what extent all EPA statutes are compatible with various aspects of sustainability. Because EPA did not request that the committee address laws pertaining to EPA or organizational and institutional aspects of the agency's operations, the committee did not examine these topics. The committee did, however, examine the benefits, where EPA has statutory authority and discretion in regulatory and nonregulatory programs, of building sustainability considerations into its administration of these statutes. The committee developed the Sustainability Framework and the Sustainability Assessment and Management approach (Figure S-1) to provide guidance to EPA on incorporating sustainability into decision making. The Sustainability Assessment and Management process is intended to be equally applicable to all types of issues, including human health and ecological risks.

The committee emphasizes in the report that the adoption of the Sustainability Framework and the application of the Sustainability Assessment and Management approach to particular EPA programs, activities, and decisions are discretionary. The committee expects that EPA will choose where to focus its attention and resources in operationalizing sustainability and in implementing its agenda and will adapt the scale and depth of the assessment according to the type of decision and its potential impact. Although it will take time and experience to incorporate sustainability broadly into EPA's culture and process, the committee anticipates that over time there will be an increasing use of the framework.

There is a broad range of sustainability activities in other federal agencies, and the committee envisions EPA working closely with these other agencies as they implement the framework. Although addressing how EPA should engage other agencies is beyond the scope of this current report, this effort will complement other programs that are addressing national and global needs for integrating sci-

---

[4] Executive Order 13514; Federal Leadership in Environmental, Energy, and Economic Performance; was signed on October 5, 2009.

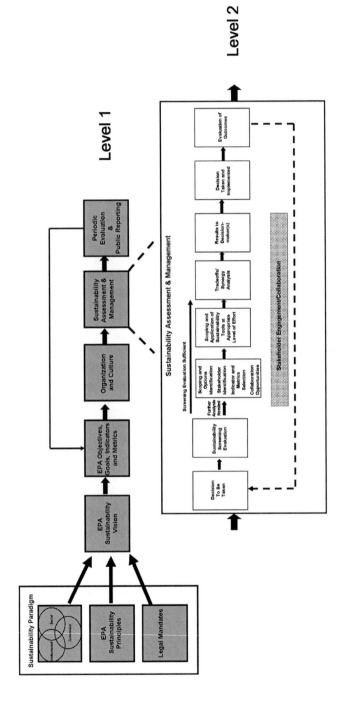

FIGURE S-1 A framework for EPA sustainability decisions.

ence and technology for sustainability, such as the National Science and Technology Council's Committee on Environment, Natural Resources, and Sustainability. In addition to engaging other agencies as EPA implements the framework, other stakeholders will also be important to engage, such as state regulators, local officials, industry, academia, community and advocacy groups, and the international community. This will better inform agency decision makers as the framework is refined to promote innovative solutions that are enriched by the growing knowledge of the interconnections of societal, environmental, and economic systems.

As with all decision making at EPA, uncertainty needs to be acknowledged and addressed, the approach needs to be transparent, and key stakeholders need to be engaged. In addition to uncertainty, tradeoff and synergy analysis is a fundamental component of the Sustainability Assessment and Management approach. The objective is to maximize social, environmental, and economic benefits of a decision and to minimize the adverse effects of conflicts among the three pillars.

The committee limited its recommendations to EPA, but these recommendations are pertinent to the concerted effort by all federal agencies and sectors of society needed to meet the challenges of achieving a sustainable future. The following recommendations were identified by the committee as key recommendations because of their importance in directly addressing the four key questions of the statement of task:

- The committee recommends that EPA adopt or adapt the comprehensive Sustainability Framework proposed in Figure S-1. The proposed Sustainability Framework requires a comprehensive approach including specific processes for incorporating sustainability into decisions and actions. As part of the framework, EPA should incorporate upfront consideration of sustainability options and analyses[5] that cover the three sustainability pillars (social, environmental, and economic), as well as trade-off considerations into its decision making. This framework was developed with the intent that EPA could apply it to any decision to which a need arose. (**Recommendation 3.1**)
- For programs, EPA should set several strategic 3–5 year breakthrough objectives[6] related to its sustainability implementation and its performance indicators and associated metrics.[7] These goals would be designed to improve performance throughout the agency by extending its

---

[5] Sustainability options and subsequent analyses refer to the range of options and the associated social (including health), environmental and economic impacts for each option along with any trade-off analyses that may have been undertaken.

[6] Commonly referred to in the business community, breakthrough objectives are goals that extend far beyond the current capabilities and experiences of an organization and require new strategies and approaches to ensure successful attainment of these goals. These objectives are generally designed to improve performance throughout an organization.

[7] Throughout the report, the committee refers to indicators and associated metrics. Indicators, in general, are measures that provide information on the state of, or change in, a system (see Box 3-3).

current capabilities and experiences and by requiring new strategies and approaches to ensure their attainment. EPA should begin periodic public sustainability reporting to transparently review its progress versus goals. **(Recommendation 3.2)**

- The committee recommends expressly including the term "health" in the social pillar to help ensure that EPA regulatory and scientific staff primarily concerned with human-health issues recognize their existing role in sustainability and recommends that EPA pay particular attention to explaining the role of human health in the social pillar, thereby ensuring that staff and stakeholders involved in the area of human health recognize that their activities are an integral part of EPA's sustainability work. Further, expressly including health in the social pillar will more clearly communicate outside of EPA the agency's role in that pillar of sustainability. **(Recommendation 3.3)**

- EPA should develop a "sustainability toolbox" that includes a suite of tools for use in the Sustainability Assessment and Management approach. Collectively, the suite of tools should have the ability to analyze present and future consequences of alternative decision options on the full range of social, environmental, and economic indicators. Application of these tools, ranging from simple to complex, should have the capability of showing the distributional impacts of options with particular reference to vulnerable or disadvantaged groups and ecosystems. **(Recommendation 4.1)**

- EPA should include risk assessment as a tool, when appropriate, as a key input into its sustainability decision making. **(Recommendation 5.1)**

- The committee recommends that EPA institute a focused program of change management to achieve the goal of incorporating sustainability into all of the agency's thinking to optimize the social, environmental, and economic benefits of its decisions; and to create a new culture among all EPA employees. **(Recommendation 6.1)**

- EPA should hire multidisciplinary professionals who are proficient in many disciplines, who have experience in the development and implementation in the sustainability assessment tools described, and who have a working knowledge in all three pillars and their application to environmental issues. The agency should hire leaders and scientists including from outside sectors to aid the agency in shifting to a more cross cutting mindset. Although EPA has existing staff in all the main areas of sustainability related fields, the agency should further facilitate collaboration among existing professional expertise to encourage dialogue and understanding of the various fields and work already being done within EPA. **(Recommendation 6.10)**

1

Introduction

The growing identification of sustainability as both a process and a goal to ensure long-term human well-being that does not threaten the continued availability of critical natural resources is based on four converging drivers. The first is the recognition that current approaches aimed at decreasing existing risks, however successful, are not capable of avoiding the complex problems in the United States and globally that threaten the planet's critical natural resources and that put current and future human generations at risk, including population growth, the widening gaps between the rich and the poor, depletion of finite natural resources, biodiversity loss, climate change, and disruption of nutrient cycles. Second, sophisticated tools are increasingly available to address the complex and challenging issues that go beyond current risk management of major threats. Third, sustainability is being used as a common approach to address broader social, environmental, and economic issues by international bodies in which the United States is an active participant. Finally, the potential economic value of sustainability to the United States is recognized to not merely decrease environmental risks but also to optimize the social and economic benefits of environmental protection.

Recognizing the importance of sustainability in meeting the agency's mission, the administrator of the U.S. Environmental Protection Agency (EPA), Lisa Jackson, formally requested that the National Research Council (NRC) undertake a study to strengthen the scientific basis for incorporating sustainability concepts into EPA's decision making. Ms. Jackson stated that incorporating sustainability into EPA's decision making is "a step toward the more effective pursuit of all of our work, including our statutory requirements, by incorporating sustainability into our foundations" (Jackson 2010). Sustainability is a fast-moving subject, for example, the White House's Council on Environmental

Quality (CEQ) recently issued guidance to the federal agencies and requested plans on incorporating sustainability into their operations. Given the time line of this NRC report, the CEQ submissions could not be factored in.

The 1969 National Environmental Policy Act (NEPA) declared that the "continuing policy of the Federal Government" is to "create and maintain conditions, under which humans and nature can exist in productive harmony, that permit fulfilling the social, economic, and other requirements of present and future generations" (42 U.S.C. 4331(a)). That policy expresses what is now described as sustainable development. Meeting the goal of sustainable development requires an integration of social, environmental, and economic policies, necessitating interdisciplinary coordination among federal agencies with varying missions to address this goal. International acceptance of sustainable development was spurred by the 1987 report of the World Commission on Environment and Development, *Our Common Future,* of which former EPA administrator, William D. Ruckelshaus, was a member (WCED 1987). Lead author of this report, Jim MacNeill, has recognized progress in institutionalizing sustainable development; however, he notes that "the need for a global transition to more sustainable forms of development, especially in the energy field, is far more urgent today than it was in 1987" when the report first issued a call for such a transition (OECD 2007). In 1992, at the United Nations Conference on Environment and Development in Rio de Janeiro, the United States and other countries endorsed a global plan of action for sustainable development and a set of principles to guide that effort (UNCED 1992a,b).

EPA'S MISSION AND ROLE

As explained more fully in Chapter 2, EPA was created in 1970 to consolidate many activities that were previously administered by several agencies. Many of the statutes administered by the new agency were intended to protect both human health and the environment from the adverse effects of pollution.[1] In spite of its name, EPA has never focused only on environmental protection.

Today, the agency's primary goals set forth in its 2011–2015 strategic plan (EPA 2010) are the following:

- Take action on climate change and improving air quality.
- Protect U.S. waters.
- Clean up communities and advance sustainable development.
- Ensure the safety of chemicals and prevent pollution.
- Enforce environmental laws.

---

[1] See www.epa.gov/lawsregs/laws for summaries of laws and Executive Orders that the EPA is charged with administering.

EPA's authorizing statutes provide numerous approaches for achieving its mission and objectives. Those approaches include the following:

- Set and enforce environmental quality standards as well as standards of performance for industrial, agricultural, and governmental sources of pollution and for producers of chemicals and pesticide products.
- Issue permits and approvals and take enforcement actions for noncompliance or do so through state environmental agencies.
- Provide grants to states and municipalities for program support and infrastructure financing.
- Provide technical guidance and assistance in both regulatory and nonregulatory programs.
- Conduct and sponsor scientific research on environmental risks and assessment, control, and measurement tools.
- Convene and collaborate with other government agencies, private corporations, academic institutions, and nongovernmental organizations on problem solving.

EPA's statutes give primary weight to protection of the environment and human health. These statutes do so for historical reasons, the most prominent being the insufficient consideration of environmental and human health protection in the past. The reasons also reflect the reality that other statutes and policies have encouraged economic development and social well-being (Friedman 2005). Nothing in this report is intended to disturb or undermine EPA's historical mission. Indeed, the committee understands part of its task as providing guidance to EPA on how it might implement its existing statutory authority to contribute more fully to a more sustainable-development trajectory for the United States.

Each agency or department of the federal government has distinct responsibilities for various social, environmental, and economic aspects of sustainability. Each agency or department can thus make a contribution to sustainability within the parameters of its existing statutory authority. The committee's purpose is to examine the benefits, where EPA has statutory authority and discretion in regulatory and nonregulatory programs, of building sustainability considerations into its administration of the statutes. The committee has not examined whether or to what extent all EPA statutes are compatible with various aspects of sustainability.

EPA's mission is based on the understanding that human health and the environment are related; what is good for the environment also tends to be good for human health, and what is good for human health tends to be good for the environment. Because EPA attempts to foster human and environmental well-being at the same time for the benefit of present and future generations, EPA's mission is consistent with sustainability. The same general drivers for sustainability described above all support an intensified effort for sustainability at EPA that would try to maximize the social, environmental, and economic benefits of

EPA's activities, and not simply reduce risk to human health and the environment. Sustainability is fully consistent with EPA's historical mission and does not undermine EPA's role of applying government regulations and policies in response to congressional mandates to protect human health and the environment and of furthering advances in environmental science and technology.

In addition, several other federal agencies have begun to integrate sustainability into their work, including the U.S. Geological Survey (USGS 2007), U.S. Department of Energy (DOE 2010), and the National Oceanic and Atmospheric Administration (Lubchenco and Sutley 2010), among others. The USGS, for example, is implementing a series of six science directions chosen to build on existing strengths: "understanding ecosystems and predicting ecosystem change; climate variability and change; energy and minerals for America's future; a national hazards, risk, and resilience assessment program; the role of environment and wildlife in human health; and a water census of the United States" (USGS 2007). USGS determined that central to their "deliberations on the content of each of the six directions was a structured framework that addressed the need to (1) identify and measure key variables, (2) map the resulting data spatially, (3) understand the fundamental natural science processes involved, (4) monitor essential variables over time, (5) predict or forecast the future course of natural science events, and (6) engage stakeholders in the use of this information for problem solving" (USGS 2007).

Additionally, the National Oceanic and Atmospheric Administration's (NOAA) 2010 Strategic Plan states the mission for the agency as "to understand and predict changes in climate, weather, oceans, and coasts, to share that knowledge and information with others, and to conserve and manage coastal and marine ecosystems and resources" and their vision is stated as "healthy ecosystems, communities, and economies that are resilient in the face of change" (NOAA 2010). In order to meet their stated mission and vision, NOAA focused on four long-term outcomes: climate adaption and mitigation; being prepared for and responding to weather-related events; healthy oceans and sustained, productive ecosystems; and resilient coastal communities and economies (NOAA 2010).

EPA has undertaken several sustainability initiatives and can claim success in developing processes leading to sustainability. It has established various sustainability programs at the program office and regional level and has adopted a sustainability research plan and highlighted sustainability in its strategic plan for 2011–2015 (EPA 2010). The agency has also examined applications in a variety of areas to better incorporate sustainability in its decision making, programs, and operations. However, as is evident in its request for the formation of this committee, EPA recognizes that to obtain the full benefits of using sustainability as a process and as a goal, the agency needs to institutionalize sustainability more broadly into its activities. Paul Anastas, assistant administrator for research and development at EPA when discussing the scientific challenge of sustainability stated, that "addressing the unsustainable trajectory of the planet has often been thought to be one that is best dealt with through government action, behavioral change, and public policy. . . . In the absence of new materials for alternative energy, food production, water purification and medicines, there are no policies that can be put into place to avoid catastrophic consequence. The magnitude and urgency of the scientific challenge cannot be understated [sic]" (Anastas 2010).

COMMITTEE'S TASK

To further strengthen the analytic and scientific basis for sustainability as it applies to human health and environmental protection, the agency requested that the NRC convene a committee under the Science and Technology for Sustainability Program to address the following questions:

- What should be the operational framework for sustainability for EPA?
- How can the EPA decision-making process rooted in the risk assessment/ risk management (RA/RM) paradigm be integrated into this new sustainability framework?
- What scientific and analytical tools are needed to support the framework?
- What expertise is needed to support the framework?

COMMITTEE'S APPROACH TO THE TASK

In response to EPA's request, the NRC appointed the Committee on Incorporating Sustainability in the U.S. Environmental Protection Agency, which prepared this report. To accomplish its task, the committee held meetings in December 2010 and February 2011. The latter meeting was a week long intensive session, which included extensive reviews of relevant literature, deliberation, and drafting of the report. In addition, data-gathering sessions that were open to the public were held during both meetings. During these public sessions, the committee heard from several EPA officials on topics ranging from the history of sustainability efforts at EPA, the principles and decision criteria for incorporating sustainability in EPA programs, and the ongoing sustainability efforts at the regional level. In addition, various state-level environmental agencies provided their perspectives on how they would use a similar sustainability framework in their programs.

Finally, the committee heard from several speakers from industry, non-governmental organizations (NGOs), and former EPA officials who provided a business perspective on sustainable development and environmental stewardship and the perspectives of various environmental groups. The committee reviewed a large body of written material on sustainability, including literature that informed the committee on how EPA could further strengthen its sustainability efforts as applicable to human health and environmental protection within the agency's decision-making process. The available data included other NRC reports, published research articles, and both U.S. and international governmental reports.

The committee also reviewed many other documents related to EPA's structure and function. The documents had the common theme of the need to balance EPA's media-specific approaches to environmental laws with its responsibilities to integrate across programs efficiently to achieve maximal environmental benefits (EPA 1992).

The committee recognized that many state and local agencies already were actively involved in advancing sustainability, and that there is also a broad range of sustainability activities in other federal agencies. The committee envisions EPA working closely with these other agencies as they implement the framework. Although addressing how EPA should engage other agencies is beyond the scope of this current report, this effort will complement other programs that are addressing national and global needs for integrating science and technology for sustainability, such as the National Science and Technology Council's Committee on Environment, Natural Resources, and Sustainability. In addition to engaging other agencies as EPA implements the framework, other stakeholders will also be important to engage, such as state regulators, local officials, industry, academia, community and advocacy groups, and the international community. This will better inform agency decision makers as the framework is refined to promote innovative solutions that are enriched by the growing knowledge of the interconnections of societal, environmental, and economic systems.

The committee's charge also did not deal with EPA's legal mandates, organizational structure, or institutional framework. There may be significant opportunities for promoting sustainability by examining these subjects but the committee was precluded from addressing these issues as they were not part of the Statement of Task. The committee also did not devote significant time to defining sustainability but used the definition from Executive Order 13514, where it is defined as

Sustainability: "to create and maintain conditions, under which humans and nature can exist in productive harmony, that permit fulfilling the social, economic, and other requirements of present and future generations" (NEPA [1969]; EO 13514[2009][2]).

The phrase, "create and maintain," captures the two senses in which the term sustainability is used by the committee in this report—as a process and as a goal. Sustainability is a process because the United States and other countries are a long way from being sustainable, and it is thus necessary to create the conditions for sustainability (NRC 1999). Sustainability is also a goal. As sustainability is achieved in particular places and contexts, it is necessary to maintain the conditions supporting it in the face of social, technological, environmental, and other changes. Although the exact nature of a sustainable society is difficult to know in advance, the basic conditions for that society (e.g., absence of large

---

[2] Executive Order 13514; Federal Leadership in Environmental, Energy, and Economic Performance; was signed on October 5, 2009.

scale poverty and environmental degradation and intergenerational responsibility) can be stated (WCED 1987).

The committee emphasizes in the report that the adoption of the Sustainability Framework and the application of the Sustainability Assessment and Management approach to particular EPA programs, activities, and decisions are discretionary. The committee expects that EPA will choose where to focus its attention and resources in operationalizing sustainability and in implementing its agenda and will adapt the scale and depth of the assessment according to the type of decision and its potential impact. Although it will take time and experience to incorporate sustainability broadly into EPA's culture and process, the committee anticipates that over time there will be an increasing use of the framework.

As with all decision making at EPA, uncertainty needs to be acknowledged and addressed, the approach needs to be transparent, and key stakeholders need to be engaged. The Sustainability Assessment and Management process is intended to be equally applicable to all types of issues, including human health and ecological risks.

The committee addressed its task by providing guidance to EPA on the processes necessary to incorporate sustainability into the agency's work but not on the specific direction EPA should take to accomplish this task. In a presentation to the committee at the first meeting, the Office of Research and Development's assistant administrator, Paul Anastas, suggested that the committee draft its recommendations in a manner following the 1983 NRC report *Risk Assessment in the Federal Government,* otherwise known as the "Red Book." The Red Book provided the agency with a framework for risk assessment and risk management (NRC 1983) that served as a driver for EPA's activities and for environmental regulations. Dr. Anastas emphasized that the sustainability equivalents of the four-box risk paradigm are needed now to serve as the critical elements of whether an action will advance or impair sustainability.

STRUCTURE OF THE REPORT

The remainder of this report is organized into seven chapters. Chapter 2 includes a history of sustainable-development activities in the United States and internationally, including documentation of the early development of sustainability concepts in U.S. environmental law. Chapter 3 describes the proposed Sustainability Framework. Chapter 4 discusses the processes and tools to support the proposed framework for sustainability at EPA. Chapter 5 provides guidance on integrating the new Sustainability Framework and the EPA decision-making process, which is rooted in the RA/RM paradigm. A discussion of cultural "change management" is provided in Chapter 6. Chapter 7 examines the relevance and utility of sustainability considerations in EPA's accomplishment of its mission. Chapters 6 and 7 also provide examples of successful sustainability initiatives.

The committee's report, although providing background information about the history and rationale for sustainability goals and processes, has not focused on precisely defining or justifying sustainability. On the basis of the Statement of Task, our focus has been on providing an operational framework that helps facilitate the further incorporation of goals for sustainability and processes in sustainability into the daily work routine of EPA.

REFERENCES

Anastas, P. 2010. Perspective on green chemistry: The most challenging synthetic transformation. Tetrahedron 66(5):1026-1027.

DOE (U.S. Department of Energy). 2010. Strategic Sustainability Performance Plan: Discovering Sustainable Solutions to Power and Secure America's Future. U.S. Department of Energy, Washington, DC [online]. Available: http://www1.eere.energy.gov/sustainability/pdfs/doe_sspp. pdf [accessed Apr. 18, 2011].

EPA (U.S. Environmental Protection Agency). 1992. Safeguarding the Future: Credible Science, Credible Decisions. EPA/600/9-91/050. Expert Panel on the Role of Science at EPA, U.S. Environmental Protection Agency, Washington, DC. March 1992.

EPA. 2010. Fiscal Year 2011-2015 EPA Strategic Vision: Achieving Our Vision. U.S. Environmental Protection Agency, Washington, DC [online]. Available: http://www.epa.gov/planandbudget/ strategicplan.html [accessed Apr. 18, 2011].

Friedman, L.M. 2005. A History of American Law, 3rd Ed. New York: Simon and Schuster.

Jackson, L.P. 2010. Remarks to the National Academy of Sciences, November 30, 2010, Washington, DC [online]. Available: http://yosemite.epa.gov/opa/admpress.nsf/8d49f7ad4bbcf4ef8525735 90040b7f6/1c893e457b3cbb25852577ec0054048c!OpenDocument [accessed Apr. 15, 2011].

Lubchenco, J., and N. Sutley. 2010. Proposed U.S. policy for ocean coast, and Great Lakes stewardship. Science 328(5985):1485-1486.

NOAA (National Oceanic and Atmospheric Administration). 2010. NOAA's Next Generation Strategic Plan. Office of Program Planning and Integration, National Oceanic and Atmospheric Administration, Silver Spring, MD. December 2010 [online]. Available: http://www.ppi.noaa. gov/wp-content/uploads/NOAA_NGSP.pdf [accessed July 11, 2011].

NRC (National Research Council). 1983. Risk Assessment in the Federal Government: Managing the Process. Washington, DC: National Academy Press.

NRC. 1999. Our Common Journey: A Transition toward Sustainability. Washington, DC: National Academy Press.

OECD (Organisation for Economic Cooperation and Development). 2007. Institutionalising Sustainable Development. Paris: OECD.

UNCED (United Nations Conference on Environment and Development). 1992a. Agenda 21. United Nations Conference on Environment and Development, Rio de Janeiro, Brazil, 3-14 June 1992 [online]. Available: http://www.cuttingthroughthematrix.com/articles/Agenda21.pdf [accessed Apr. 18, 2011].

UNCED. 1992b. Report of the United Nations Conference on Environment and Development, Rio de Janeiro, 3-14 June 1992, Annex I. Rio Declaration on Environment and Development. A/CONF.151/26 (Vol. I). United Nations General Assembly [online]. Available: http://www. un.org/documents/ga/conf151/aconf15126-1annex1.htm [accessed Apr. 15, 2011].

USGS (U.S. Geological Survey). 2007. Facing Tomorrow's Challenges: U.S. Geological Survey Science in the Decade 2007-2017. Circular 1309. U.S. Department of the Interior, U.S. Geological Survey, Reston, VA [online]. Available: http://pubs.usgs.gov/circ/2007/1309/pdf/C1309.pdf [accessed Apr. 18, 2011].

WCED (United Nations World Commission on Environment and Development). 1987. Our Common Future. Oxford: Oxford University Press.

2

History of Sustainability

Sustainability is based on a simple and long-recognized factual premise: Everything that humans require for their survival and well-being depends, directly or indirectly, on the natural environment (Marsh 1864). The environment provides the air we breathe, the water we drink, and the food we eat. It defines in fundamental ways the communities in which we live and is the source for renewable and nonrenewable resources on which civilization depends. Our health and well-being, our economy, and our security all require a high quality environment.

When we act on that understanding, we tend to prosper; when we do not, we suffer. For example, the Dust Bowl of the 1930s occurred because wheat farmers were encouraged by the federal government to plow up sod across large areas of the high plains in Texas and Oklahoma at a time when precipitation was more plentiful. When customarily dry conditions recurred, huge dust storms swept across the unprotected landscape, making farming impractical and life much more difficult and hazardous due to dust pneumonia. Soil conservation practices, including crop rotation and fallowing land, were introduced on a large-scale basis afterward, and the Dust Bowl has not recurred (Egan 2006). Nonetheless, aquifer depletion, climate change, and unsustainable farming practices all render the Great Plains increasingly vulnerable to severe drought (Adler 2010).

This chapter provides a brief history of the concept of sustainable development or sustainability. Although the Committee on Incorporating Sustainability in the U.S. Environmental Protection Agency (EPA) was not specifically tasked to provide an historical overview, this history is useful in providing context for the rest of the report. Conceptually, sustainable development emerged as a result of significant concerns about the unintended social, environmental, and economic consequences of rapid population growth, economic growth, and consumption of natural resources. This history has three overlapping story lines, as more fully explained below. The first occurred in the United States as a conservation movement, which developed from the recognition that our taming of the wilderness was destroying much of what we valued as part of the U.S. culture—a recognition that led to conservation laws which began to emerge in the late nineteenth century. The second was based on the realization that some of the chemical and physical agents increasingly released into the environment because of industrial development were harmful to people and the environment—a realization that led to such events as the original Earth Day and the formation of EPA in 1970 and the ensuing media and pollutant-based environmental laws.

The third story line is based upon the perception that population growth and consumption are challenging the ability of Earth's ecosystems to provide for future generations and that the response to this challenge requires more than "place-based" (see Appendix C) conservation or the control of environmental pollutants. The institutionalization of this began with a series of international conferences and agreements that were—to a very large degree—based on and inspired by actions that were already under way in the United States. Although formal international endorsement of sustainable development occurred at the United Nations Conference on Environment and Development (UNCED or Earth Summit) in Rio de Janeiro in 1992, many of its underlying concepts and principles had long been recognized in U.S. law and policy. Since the Earth Summit, the most successful U.S. efforts have been in response to stakeholder or constituent demand. However, in contrast to the United States, the third story line also contains the explicit and strategic use of the concept of sustainable development in other developed countries.

CONSERVATION IN THE UNITED STATES

The conservation and preservation movements in the United States—and the laws that were enacted in response to these—represented an effort to reconcile economic development with the protection of the environment by ensuring the availability of natural resources for the benefit of both present and future generations (Van Hise 1927, Fox 1981). It was also a response to the destruction of native virgin forests by logging and conversion to agriculture, as well as to the extinction of species such as the passenger pigeon and the near extinction of the American bison (more popularly known as the buffalo). As the movement evolved over the late nineteenth and early twentieth centuries, its objectives included protection of forests, water, soils, public lands, and wildlife (Beatty 1952, Hays 1959, Reiger 1975, Norse 2005). Early fish biologists and ecologists also played an important role in advancing the concepts and methods related to sustainable fish consumption and harvesting and sustainable ecosystems. There was also an understanding among American's leading conservationists that human well-being relied on all natural resources. Gifford Pinchot, the first chief of the U.S. Forest Service, wrote that "our unexampled wealth and well-being are directly due to the superb natural resources of our country" (Pinchot 1910). Pinchot later added that the first purpose of conservation policy is "wisely to use, protect, preserve, and renew the natural resources of the earth" (Pinchot 1947). Conservation was successful due to the vision of leading conservationists such as Pinchot and John Muir; however, the movement also required political leadership to gain traction. In the early 1900s, Theodore Roosevelt, the "conservation president," signed legislation establishing five national parks, and created or expanded many national forests, wildlife preserves, and other conservation areas (Brinkley 2009). "The conservation of our natural resources and their proper use constitute the fundamental problem which underlies almost every other problem of our National life," Roosevelt told Congress. "We must maintain for our civilization the adequate material basis without which that civilization can not exist. We must show foresight, we must look ahead" (Roosevelt 1907). Other conservation laws and programs require or encourage greater efficiency in the use of natural resources, and still others impose limits on harvesting natural resources so that those resources will be able to regenerate or reproduce for use in the future (Hays 1959; Leopold 1986).

ENVIRONMENTAL PROTECTION IN THE UNITED STATES

The environmental movement in the United States, which is broad in scope, responded to growing industrialization, population, and pollution, as well as to resource exploitation (Lazarus 2004). It was motivated by a public desire for higher quality of life and well-being, improved human health, and long-term protection of ecosystems (Hays 1987). A major issue is adverse effects of pollutants, pesticides, and chemicals on humans and the environment. For example, at least 20 people died and thousands were sickened in 1948 in Donora, Pennsylvania, during an episode of industrial air pollution. In 1962, Rachel Carson published *Silent Spring,* which described the potential impact of pesticides on birds and animals and suggested that humans were also being harmed (Carson 1962). Public perception of dirty air and rivers that were no longer suitable for swimming or fishing, and a landscape littered with industrial waste were driving forces in the development of media-specific laws and in the formation of EPA. There was also concern that the federal government was often supporting environmentally damaging economic development in the form of federal dams, highway projects, stream-channelization and flood-control projects, and other activities that had unintended adverse environmental effects (Andrews 2006).

The first major federal environmental law is the National Environmental Policy Act of 1969 (NEPA). In declaring a national policy "to create and maintain conditions under which man and nature can exist in productive harmony, and fulfill the social, economic, and other requirements of present and future generations of Americans" (42 U.S.C. § 4331(a)), Congress provided a statutory foundation for sustainability within the EPA. By its very nature, NEPA emphasizes the importance of sustainability. This provision is particularly true because Congress then stated that "the continuing responsibility of the Federal Government" is to, among other things, "fulfill the responsibilities of each generation as trustee of the environment for succeeding generations" (42 U.S.C. § 4331(b)(1)). NEPA then states that this and other similar responsibilities are in addition to existing grants of agency authority: "The policies and goals set forth in this Act are supplementary to those set forth in existing authorizations of Federal agencies" (42 U.S.C. § 4335; ELI 1995).

Beyond its declaration of policy, NEPA requires that federal agencies prepare an environmental impact statement before taking a major action "significantly affecting the quality of the human environment" (42 U.S.C. § 4332 (c)). The statement is to include both a description of the environmental effects of the proposed action as well as alternatives to that action. In this way, NEPA requires federal agencies take into account environmental considerations into their decision-making processes (42 U.S.C. § 4332 (c)).[1]

Through the 1950s and 1960s, Congress passed legislation concerning air quality, water quality, and other environmental problems. Beginning in 1970, however, it overhauled these prior laws to impose limits and permitting requirements to protect air quality (Clean Air Amendments of 1970) and water quality (Federal Water Pollution Control Amendments of 1972 [PL 92-500]), to protect drinking water (Safe Drinking Water Act of 1974), and to prevent and control adverse effects from the improper disposal of solid and hazardous waste (Resource Conservation and Recovery Act [RCRA] of 1976). These statutes used a "cooperative federalism" approach in which the federal government sets standards and states are given substantial financial support to enforce and implement these requirements.

In 1980, in response to risks presented by sites where hazardous substances had been improperly disposed, Congress adopted the Comprehensive Environmental Response, Compensation, and Liability Act (CERCLA), which established Superfund. This act imposes liability on certain parties for conditions on these sites and establishes a process for their remediation. Additionally, the Federal Insecticide, Fungicide, and Rodenticide Act (FIFRA), which was first passed in 1947, now mandates that EPA "regulate the use and sale of pesticides to protect human health and preserve the environment" (EPA 2010).

In the late 1960s, at the beginning of the modern environmental era, federal responsibility for environmental protection was divided among many federal agencies, including the U.S. Department of the Interior (water quality), the U.S.

---

[1] Both Congress and the courts have decided that the environmental impact statement requirements of NEPA are generally inapplicable to EPA decisions, in no small part because the statutes EPA administers contain information gathering and analytical requirements that are considered the "functional equivalent" of an environmental impact statement (Rodgers 1994, 1999, § 9.5(D)(2)). That exemption does not appear to apply to other provisions of NEPA, however.

Department of Agriculture (pesticides), and the U.S. Department of Health, Education, and Welfare (air quality). An advisory council identified the government organization at that time as an impediment to effectively addressing "the environmental crisis, noting that "many agency missions . . . are designed primarily along media lines. . . . Yet the sources of air, water, and land pollution are interrelated and often interchangeable." The advisory council added that "some pollutants—chemicals, radiation, pesticides—appear in all media. Successful interdiction now requires the coordinated efforts of a variety of separate agencies and departments. The result is a blurring of focus, and a certain Federally-sponsored irrationality" (PACEO 1970).

In response to the advisory council, President Richard Nixon in 1970 created the EPA by a reorganization plan that transferred to the new agency a variety of environmental functions from four federal agencies (Reorganization Plan No. 3 of 1970, codified at 5A U.S.C.). The agency's overall mission, then and now, is to protect human health and the environment.

EPA is the primary federal agency responsible for administering most of the major environmental statutes, including the Clean Air Act, the Clean Water Act, RCRA, the Safe Drinking Water Act, and CERCLA. However, EPA is not the only agency with environmental responsibilities. The U.S. Department of the Interior, for example, is the federal agency with primary responsibility for administering the Endangered Species Act (1973) and the Surface Mining Control and Reclamation Act (1977). The U.S. Department of Transportation and the U.S. Department of Energy (DOE) have significant responsibilities for federal energy efficiency, operation, and cleanup at DOE sites nationwide and for meeting conservation requirements, which affect the environment in a variety of ways. The U.S. Army Corps of Engineers Civil Works Program administers, in coordination with the EPA, the issuance of permits under section 404 of the 1972 Federal Water Pollution Act amendments, which controls development in wetlands across the nation.

SUSTAINABLE DEVELOPMENT

Although sustainable development was formally endorsed at an international conference in 1992, it was supported by the United States and is based to a significant degree on U.S. law and experience. Since that time, the United States has approached sustainable development in a manner that is somewhat different from other countries, particularly developed countries, as discussed below.

Sustainable Development at the International Level

At the end of World War II, the United States led an effort to create a system of international agreements and institutions based on two pillars—economic development as well as social development or human rights—that are predicated on a foundation of peace and security. These key elements formed the basis of the concept of development as it was formally understood by the international community (Dernbach 1998; Schlesinger 2003; Borgwardt 2005). The ultimate aims of development are human well-being, quality of life, freedom, and opportunity (WCED 1987; Sen 1999; Sarkar 2009; De Feyter 2001).

Development has worked well in many ways. Living standards have increased around the world, the global economy has grown, and people are living longer (UNDP 1999). Development has also caused growing problems of resource exploitation and pollution around the world. These concerns led to the creation of the Environment Committee in the Organisation for Economic Co-operation and Development (OECD) (OECD 2001), which held its inaugural meeting under an American chairperson in November 1970. In 1972, in Stockholm, the United Nations (UN) Conference on the Human Environment agreed to establish the UN Environmental Programme (UNEP) (UNEP 2011a). This conference did not, however, provide a framework for reconciling development with environmental protection.

In the 1970s and 1980s, however, it became increasingly clear that the interrelated issues of widespread poverty and growing environmental degradation around the world were not being effectively addressed and that the development model needed to be modified.

In 1980, the International Union for Conservation of Nature (IUCN) published its *World Conservation Strategy: Living Resource Conservation for Sustainable Development* (IUCN 1980). The strategy represented the "integration of conservation and development" in the form of "sustainable development." It defined conservation as the "management of human use of the biosphere so that it may yield the greatest sustainable benefit to present generations while maintaining its potential to meet the needs and aspirations of future generations." The IUCN acknowledged the difficulty of merging the two concepts: "Conservation and development have so seldom been combined that they often appear—and are sometimes represented as being—incompatible." It nonetheless concluded that "integration of conservation and development" is needed to "ensure that modifications to the planet do indeed secure the survival and well-being of all people."

The World Commission on Environment and Development (known as the Brundtland Commission, after its chair, Gro Harlem Brundtland), adopted this approach in its seminal 1987 report, *Our Common Future* (WCED 1987). The commission was created by a UN General Assembly Resolution in 1983 to "propose long-term environmental strategies for achieving sustainable development to the year 2000 and beyond" (UNGA 1983). The report, which described "a threatened future," provided the iconic definition of sustainable development: "development that meets the needs of the present without compromising the ability of future generations to meet their own needs." The Commission also called upon the UN General Assembly to transform its report into a global action plan for sustainable development.

The nations of the world did precisely that at the 1992 UN Conference on Environment and Development, or "Earth Summit," in Rio de Janeiro (scheduled to coincide with the twentieth anniversary of the Stockholm Conference). These nations, including the United States, endorsed a global sustainable development action plan, known as Agenda 21 (UNCED 1992a), and a set of 27 principles for sustainable-development, called the Rio Declaration (UNCED 1992b). Together, these agreements modify the definition of development by adding a third pillar—environmental protection and restoration—to the economic and social pillars of development, and is also known as the "Triple Bottom Line" approach in the corporate sector. Sustainable development has the same ultimate aims as development—human well-being, quality of life, freedom, and opportunity. It also requires a foundation of peace and security (UNCED 1992a,b; Dernbach 1998; UN 2002).

The principles of the Rio Declaration are generally recognized as foundational to global sustainability. Many of the principles are similar to those contained in U.S. conservation and environmental law. They include the following:

- "Human beings are at the center of concerns for sustainable development" (Principle 1). This principle makes clear that human well-being and quality of life is the objective of sustainability. This declaration is similar to that contained in NEPA. Achieving sustainable development requires recognizing the need to balance the conservation of resources while protecting humans from the uncertainties of nature.
- "In order to achieve sustainable development, environmental protection shall constitute an integral part of the development process and cannot be considered in isolation from it" (Principle 4). This principle—integrated decision making—is the fundamental action principle of sustainability because it integrates the social, environmental, and economic decision making on issues, rather than considering the environmental issues separately (Dernbach 2003). This principle is reflected in different ways in each U.S. conservation and environmental law.
- Precautionary approach. "Where there are threats of serious or irreversible damage, lack of full scientific certainty shall not be used as a reason for postponing cost-effective measures to prevent environmental degradation" (Principle 15). The U.S. Clean Air Act and other environmental laws enable the adoption of standards based on the possibility of harm rather than complete certainty (Ashford and Caldart 2008). (The relation between sustainability and precaution also has been considered by O'Riordan and Cameron 1994, as well as others.)
- Intergenerational equity. The Rio Declaration's acknowledgement of the need "to equitably meet developmental and environmental needs of present and future generations" (Principle 3) is reflected expressly in NEPA and implicitly in nearly all U.S. laws related to the environment.

The sustainability literature has emphasized the need for social justice and equity, particularly in the global context. Dernbach (2002) notes the link between these concepts: "Poor people in developed and developing countries tend to be exposed to the worst environmental conditions . . . without efforts to reduce poverty and environmental degradation for the present generation, it will be difficult to ensure that future generations will have the same access to the same quality of environment or developmental conditions as the present generation."

- Internalization of environmental costs (Principle 16). The "approach that the polluter should, in principle, bear the cost of pollution" is reflected in varying degrees throughout U.S. environmental law. (At the international level, the "polluter pays principle" had earlier been adopted by the OECD Council on May 26,1972 as part of the OECD *Guiding Principles Concerning the International Economic Aspects of Environmental Policies*).
- Public participation in decision making (Principle 10).
- "Environmental issues are best handled with participation of all concerned citizens, at the relevant level." The Rio Declaration also supports public access to information as well as justice. U.S. environmental law is based on a variety of opportunities for public participation (ELI 1991).

The commitments to sustainable development made at UNCED have been essentially reaffirmed, with differing levels of emphasis, in a variety of meetings and conferences since 1992. Box 2-1 identifies some of the key meetings. Sustainable-development concepts have also been incorporated into a variety of international treaties, including the UN Framework Convention on Climate Change and the Convention on Biological Diversity, both of which were opened

---

BOX 2-1
International Sustainable Development Conferences

Several commitments and conferences related to sustainable development are of note:

- Agenda 21, Rio Declaration on Environment and Development (UNCED 1992a,b)
- Programme for the Further Implementation of Agenda 21 (UN 1997)
- Millennium Declaration (UN 2000)
- Johannesburg Plan of Implementation, World Summit on Sustainable Development (UN 2002)
- UN Conference on Sustainable Development, Rio de Janeiro, June 4-6, 2012 (UNCSD 2011)

for signature at the Earth Summit. As the twentieth anniversary of the Earth Summit approaches, sustainable development has become a key part of the generally recognized international framework for maintaining and improving the human condition (UNGA 2010). Moreover, apart from international conferences and declarations, a great many sustainability activities are occurring throughout the world, and particularly by nongovernmental organizations and the private sector who often refer to sustainability as the "triple bottom line" (Hawken 2007, WBCSD 2011).

The long-term importance of this framework is underscored by a 1999 NRC report, *Our Common Journey,* which said that it could take at least two generations (until 2050) to achieve a *transition* to sustainability (NRC 1999). The recommended primary goals of this transition "should be to meet the needs of a much larger but stabilizing human population, to sustain the life support systems of the planet, and to substantially reduce hunger and poverty" (p.4).

The framework also requires new forms of knowledge. Sustainability science has arisen as an emerging field that is problem-driven and interdisciplinary and sets a goal of "creating and applying knowledge in support of decision making for sustainable development" (Clark and Dickson 2003, Clark 2007). By drawing on multiple disciplines, such as law, engineering, and social and natural sciences, sustainability science is "defined by the problems it addresses rather than the disciplines it uses" (Clark 2007). Fiksel et al. (2009) noted that "EPA must continue to use science to fulfill its mandate" to protect human health and the environment and also to "use sustainability science to move beyond the current regulatory framework and to develop a more integrated systems-based approach to address challenges of this new century."

Sustainable Development Outside the United States

Agenda 21 and the Rio Declaration were not simply agreements about sustainability ideas; they were also agreements to achieve sustainability. Certain subsequent actions and experiences of countries and regions outside the United States, particularly those of the European Union (EU), are thus worth noting briefly. Even a brief review suggests that many countries, including developed countries, tend to address sustainable development as a policy objective or framework. As of 2009, 106 UN member countries were implementing national sustainable-development strategies (UNESA 2011). The EU's sustainable-development strategy is particularly relevant to the United States. The EU's sustainable-development strategy was first adopted in 2001 and then renewed in 2006 (CEU 2006a,b). Its basic aims are to exploit "the mutually reinforcing elements of economic, social and environment policy" and to avoid or minimize trade-offs among goals (CEC 2005, p.4). "Sustainable development offers the [EU] a positive long-term vision of a society that is more prosperous and more just, and which promises a cleaner, safer, healthier environment—a society which delivers a better quality of life for us, for our children, and for our grandchildren" (CEC 2001, p.2). The strategy identifies the following areas as priorities and contains specific measures to address them: climate change and clean energy; sustainable transport; sustainable consumption and production; conservation and management of natural resources; public health; social inclusion, demography, and migration; and global poverty and sustainable development challenges (CEU 2006b). In a 2009 review of progress in implementing its sustainable-development strategy, the European Commission said that "the EU has demonstrated its clear commitment to sustainable development and has successfully mainstreamed this sustainability dimension into many policy fields" (CEC 2009, p.3). The review added that integration of policy objectives is "improving the cost-efficiency of policy decisions" (p.3), and noted progress in developing "a low-carbon and resource-efficient economy" (p.3), which it said would be a key to economic recovery. Still, "unsustainable trends persist and the EU still needs to intensify its efforts" (p.15). The European Commission issues a biennial report that monitors the EU's progress in implementing its sustainable-development strategy; the most recent report was issued in 2009 (Box 2-2) (CEC 2009).

The OECD, which is composed of 34 of the world's most highly developed countries, including the United States, creates and analyzes information and trends concerning the environment and sustainable development and provides opportunities for relevant government officials in OECD countries to meet and share information and ideas concerning good policy practice in the areas of environment and sustainable development and to adopt internationally binding agreements in some of them, notably, chemical safety and hazardous waste. For example, in 1989, in the context of freshwater use, the OECD developed the

---

BOX 2-2
2009 Monitoring Report of EU
Sustainable-Development Strategy

The European Commission's report uses more than 100 indicators but identifies 11 "headline indicators" to provide an "overall picture of whether the EU has achieved progress toward sustainable development in terms of the objectives and targets" identified in the strategy. Progress on two indicators—gross domestic product (GDP) per capita and resource productivity—was identified as clearly favorable, using the symbol of a shining sun. For other indicators, there was no or moderately favorable change, indicated by the symbol of a sun obscured by clouds. These indicators included energy consumption of transportation in relation to GDP growth, healthy life years, the employment rate of older workers, and the abundance of common birds. Unfavorable trends included greenhouse gas emissions (moderately unfavorable, symbolized by clouds) and conservation of fish stocks (clearly unfavorable, symbolized by clouds with lightning) (CEU 2009).

"user pays principle," a concept of pricing natural-resource use to "at least cover the opportunity costs of these services: the capital, operation, maintenance, and environmental costs" (Ruffing 2010). According to the OECD, "these opportunity costs should reflect the long-run incremental costs to the community of satisfying marginal demand" (Ruffing 2010). Such a charging system is usually known "as long-run marginal social cost pricing" (Ruffing 2010).

Sustainable Development in the United States

Many of the key principles and concepts in sustainable development are rooted in, or similar to, concepts in U.S. conservation and environmental law. Generally, U.S. conservation and environmental law has advanced sustainability in some areas. Nonetheless, the United States has not used a national strategy or sustainability "indicators" (see Appendix C), and a great deal more needs to be done to achieve sustainability in the United States.

U.S. environmental and conservation laws are related to all three pillars of sustainability, not just the environmental pillar. The laws have at least nine purposes, including protection of human health, preservation for aesthetics or recreation, biocentrism, sustainability of the resource base, environmental justice, efficiency, pursuit of scientific knowledge and technology, intergenerational equity, and community stability (Campbell-Mohn 1993). The purposes of environmental and conservation laws are not limited to environmental protection; these laws also have social and economic development goals and effects. There is inherent difficulty in labeling any of these purposes as strictly social, environmental, or economic. Protection of human health, for instance, can be understood as environmental because it primarily concerns protection from pollutants, waste, and chemicals that are emitted or discharged into the environment. Yet human health protection can also be understood as social and economic because it involves humans rather than the environment and other species and often involves equity issues, such as the benefits accrual to parties different from those burdened with significant risk. On the other hand, biocentrism, which "seeks to preserve natural systems because they have inherent value beyond their usefulness to humans," and which is only weakly reflected in U.S. environmental law (Campbell-Mohn 1993), along with ecological risk, fit primarily in the environmental pillar.

In addition, cost-effective programs have been established in the United States that resulted in lower pollutant emissions. For example, in 1990 Congress amended the Clean Air Act to reduce sulfur dioxide emissions from large coal-fired power plants by 50 percent over 10 years (104 Stat. 2468, P.L. 101-549). The act used a cap-and-trade program to achieve that result. Under this program, plants with lower control costs that reduce their emissions beyond legal requirements are allowed to "trade" their excess reductions to plants with higher control costs, thus enabling a cost-effective way to achieve the emission limit. The program cost only 20 to 30 percent of projected expenditures (EDF 2011).

In addition, a 2003 Office of Management and Budget (OMB) study found that this program accounted for over $70 billion annually in quantified human health benefits—the largest of any major federal regulatory program implemented in the last 10 years (OMB 2003).

Finally, environmental and conservation laws have also had the effect of fostering sustainability in the United States. The air is cleaner and more healthful to breathe, our rivers and lakes are cleaner, and waste is much better managed, even as the economy has grown (EPA 2008). This development means that EPA has fostered sustainability to some degree through its implementation of these laws. In spite of the similarities between U.S. environmental law and sustainable development, there are some important differences. Most obviously, sustainable development is a normative conceptual framework that is broader than the sum of U.S. environmental and conservation laws. Sustainable development also raises questions that are not fully or directly addressed in U.S. law or policy, including how to define and control unsustainable patterns of production and consumption and how to encourage the development of sustainable communities, biodiversity protection, clean energy, environmentally sustainable economic development, and climate change controls. Each of these questions needs to be addressed across government agencies.

During President Clinton's Administration, the United States took a step in the direction of a national effort on behalf of sustainability with the President's Council on Sustainable Development (PCSD). Created by an executive order in 1993 and terminated by another executive order in 1999, the council issued a series of reports containing recommendations for sustainability. Its primary report was *Sustainable America: A New Consensus for Prosperity, Opportunity, and a Healthy Environment for the Future* (PCSD 1996). It stated, "A sustainable United States will have a growing economy that provides equitable opportunities for satisfying livelihoods and a safe, healthy, high quality of life for current and future generations" (p.iv). None of the PCSD's reports, however, constituted a national strategy or provided for any continuing effort on behalf of sustainability at the national level; nor did the federal government follow up on many of the report's recommendations. Since the elimination of the PCSD, there has been no federal governmental body or organization tasked with determining or implementing a coordinated sustainable-development policy for the United States.

According to the most recent OECD review for the United States, the country was well above the OECD average for per capita water use and per capita carbon dioxide emissions. U.S. emissions of carbon dioxide, sulfur oxides, and nitrogen oxides were also well above the OECD average per unit of gross domestic product (GDP) (OECD 2005). The report also stated, "Decoupling of environmental pressure from economic growth has been achieved in some areas, but the United States still faces challenges with respect to high energy and water intensities, environmental health risks, marine habitat conservation and maintenance of biodiversity" (p.1).

In addition, the Environmental Law Institute (ELI) has published a series of reports since 1997 describing and assessing U.S. sustainability efforts, and making recommendations (Dernbach 1997, 2002, 2009). These reports indicate that U.S. progress is modest at best. The 2009 report nonetheless identified six areas where considerable progress is occurring. These areas are local governance, brownfield redevelopment, business and industry, colleges and universities, kindergarten through 12th grade education, and religious organizations. A common characteristic of these areas is that their efforts are driven by the threats of climate change (NRC 2010a, IPCC 2007), the global deterioration of ecosystems (MEA 2005), and the availability of more sustainable ways of approaching these and other issues (e.g., NRC 2010b; TEEB 2010). Another common characteristic of these six areas is that their efforts are driven by their members, customers, citizens, and stakeholders. For corporations, other sustainability drivers include cost savings, competitive advantage, economic opportunity, and consumer demand, not simply avoidance of government regulation (Feldman 2009; Porter and Kramer 2011). For communities, other sustainability drivers are cost savings, reducing demand on utilities and infrastructure, and a desire to have more attractive places to live and work (Mazmanian and Kraft 2009, Weiss 2009). In all six of the areas, sustainability practitioners are learning what works and what does not work from their peers, are using new communication technologies to share information more rapidly, and are engaging in steadily more ambitious and effective efforts to maximize environmental, economic, and social value.

Sustainability efforts in the United States are also increasingly affected by three facts. First, the sustainability literature has made it clear that environmental law and regulation provide only a set of legal approaches for sustainability and that other approaches and incentives (e.g., subsidies, tax law, economic development law, and private certification) also have an important role to play (UNCED 1992a, Richardson and Wood 2006). The other tools have come into greater focus as ways, for example, to foster more sustainable communities (Fitzgerald 2010). Somewhat similarly, the U.S. Department of Energy's (DOE's) Better Buildings Initiative is using grants to state and local governments to help develop an economic infrastructure that will make it easier for homeowners and business owners to do cost-effective energy efficiency upgrades and retrofits of existing buildings (DOE 2011).

Second, the economic recession that began in 2008 has helped reframe the sustainability dialogue to some degree in terms of "green jobs" and "green business." In June 2009, the OECD governments, including the United States, adopted a "Declaration on Green Growth," recognizing that "a number of well-targeted policy instruments" (p.2) encouraging green investment could help enable a short-term economic recovery and create a more sustainable infrastructure for the long term. The OECD also called for the development of "a Green Growth Strategy in order to achieve economic recovery and environmentally and socially sustainable economic growth" (OECD 2009, p.3). The Green Growth Strategy was subsequently submitted to the Meeting of the OECD Council at Ministerial Level, 25-26 May 2011 for endorsement (OECD 2011). In its 2011 report, *Towards a Green Economy,* the UN Environment Program advocates a shift in investment in key sectors (e.g., agriculture and energy) and suggests policies such as "reduction or elimination of environmentally harmful subsidies" (p.9) to achieve this shift (UNEP 2011b). The clean-energy sector in particular is seen by many, particularly at the state and local level, as a source of economic and job creation opportunity for the United States (Byrne et al. 2007). One of the two themes for the conference to be held on the twentieth anniversary of the Earth Summit, the UN Conference on Sustainable Development in Rio de Janeiro in June 2012, is "a green economy in the context of sustainable development and poverty eradication." (The other theme is the institutional framework for sustainable development [UNCSD 2011]).

The third factor affecting U.S. sustainability decisions is global competitiveness. The globalization of economic activity and the accompanying result of emergent global scale problems (e.g., biodiversity, climate change, risk of pandemics), limitations of current institutional approaches at the global, national, regional and local levels, and the evolution of global, middle-class consumer values in major emerging markets help explain why sustainability has emerged as such a powerful challenge and opportunity for EPA and other institutions in the United States. China is increasingly seen as a major, even dominating, economic competitor in renewable energy and certain other forms of clean energy. The reasons for China's competitiveness include the government's support for such energy businesses, the inexpensive labor costs, and the large size of the Chinese market. China provides an additional reason for the United States to more aggressively pursue development of clean-energy technologies and sustainability. The EU also emphasizes the value of sustainability to its economic competitiveness. In its 2006 Annex to the "Renewed EU Sustainable Development Strategy," the economy is listed first. (Sustainability "promotes dynamic economy with full employment, and a high level of education, health protection, social and territorial cohesion and environmental protection" [CEU 2006b, p.2]).

Sustainability in This Report

The 2009 Executive Order [EO 13514] applies a definition of sustainability that is drawn from NEPA: "to create and maintain conditions, under which humans and nature can exist in productive harmony, that permit fulfilling the social, economic, and other requirements of present and future generations." This report also uses that definition.

The phrase, "create and maintain," captures the two senses in which the term sustainability is used by the committee in this report—as a process and as a goal. Sustainability is a process because the United States and other countries are a long way from being sustainable, and it is thus necessary to create the conditions for sustainability (NRC 1999). Sustainability is also a goal. As sustainability is achieved in particular places and contexts, it is necessary to maintain the conditions supporting it in the face of social, technological, environmental, and other changes. Although the exact nature of a sustainable society is difficult to know in advance, the basic conditions for that society (e.g., absence of large scale poverty and environmental degradation and intergenerational responsibility) can be stated (WCED 1987).

Looking Ahead

Thus, sustainability is gaining increasing recognition as a useful framework for addressing otherwise intractable problems. The framework can be applied at any scale of governance, in nearly any situation, and anywhere in the world. Although it was created to address serious problems—growing global environmental degradation and poverty—sustainability provides a way to address these problems in a way that can also create even greater opportunity.

FINDINGS AND RECOMMENDATIONS

2.1 Finding: EPA's historical mission is to protect human health and the environment (p.19).

2.1 Recommendation: EPA should carry out its historical mission to protect human health and the environment in a manner that optimizes the social, environmental, and economic benefits of its decisions.

REFERENCES

Adler, R.W. 2010. Drought, sustainability, and the law. Sustainability 2(7):2176-2196.

Andrews, R.N.L. 2006. Managing the Environment, Managing Ourselves: A History of American Environmental Policy, 2nd Ed. New Haven: Yale University Press.

Ashford, N.A., and C.C. Caldart. 2008. Environmental and occupational health protection law. Pp. 390-401 in International Encyclopedia of Public Health, Vol. 2, 1st Ed. Amsterdam: Elsevier.

Beatty, R.O. 1952. The conservation movement. Ann. Am. Acad. Polit. Soc. Sci. 281(May):10-19.

Borgwardt, E. 2005. A New Deal for the World: America's Vision for Human Rights. Cambridge: Belknap Press.

Brinkley, D. 2009. The Wilderness Warrior: Theodore Roosevelt and the Crusade for America. New York: HarperCollins.

Byrne, J., K. Hughes, W. Rickerson, and L. Kurdgelashvili. 2007. American policy conflict in the greenhouse: Divergent trends in federal, regional, state, and local green energy and climate change policy. Energ. Policy 35(9):4555-4573.

Campbell-Mohn, C., ed. 1993. Sustainable Environmental Law: Integrating Natural Resource and Pollution Abatement Law from Resources to Recovery. St Paul, MN: West Publishing Co.

Carson, R. 1962. Silent Spring. New York: Houghton Mifflin.

CEC (Commission of the European Communities). 2001. A Sustainable Europe for a Better World: A European Strategy for Sustainable Development. Communication from the Commission, Brussels, May 15, 2001[online]. Available: http://eur-lex.europa.eu/LexUriServ/site/en/com/2001/com2001_0264en01.pdf [accessed Apr. 18, 2011].

CEC. 2005. On the Review of the Sustainable Development Strategy: A Platform for Action. Communication from the Commission to the Council, and the European Parliament, Brussels, December 12, 2005 [online]. Available: http://www.central2013.eu/fileadmin/user_upload/Downloads/Document_Centre/OP_Resources/COM__2005__658_Sust_Dev.pdf [accessed Apr. 18, 2011].

CEC. 2009. Mainstreaming Sustainable Development into EU Policies: 2009 Review of the European Union Strategy for Sustainable Development. Communication from the Commission to the European Parliament, the Council, the European Economic and Social Committee and Committee of the Regions, Brussels, July 24, 2009 [online]. Available: http://ec.europa.eu/sustainable/docs/com_2009_400_en.pdf [accessed Apr. 18, 2011].

CEU (Council of the European Union). 2006a. Presidency Conclusions of the Brussels European Council (15/16 June 2006). 100633/1/06. Council of the European Union, Brussels, July 17, 2006 [online]. Available: http://www.consilium.europa.eu/ueDocs/cms_Data/docs/pressData/en/ec/90111.pdf [accessed Apr. 18, 2011].

CEU. 2006b. Review of the EU Sustainable Development Strategy (EU-SDS)-Renewed Strategy. 10117/06. Council of the European Union Brussels, June 9, 2006 [online]. Available: http://register.consilium.europa.eu/pdf/en/06/st10/st10117.en06.pdf [accessed Apr. 18, 2011].

CEU. 2009. 2009 Review of the European Union Strategy for Sustainable Development. Council of the European Union, Brussels, December 1, 2009 [online]. Available: http://register.consilium.europa.eu/pdf/en/09/st16/st16818.en09.pdf [accessed July 12, 2011].

Clark, W.C. 2007. Sustainability science: A room of its own. Proc. Natl. Acad. Sci. U.S.A. 104(6):1737-1738.

Clark, W.C., and N.M. Dickson. 2003. Sustainability science: The emerging research program. Proc. Natl. Acad. Sci. U.S.A. 100(14):8059-8061.

De Feyter, K. 2001. World Development Law: Sharing Responsibility for Development. Antwerp: Intersentia.

Dernbach, J.C. 1997. U.S. Adherence to its Agenda 21 Commitments: A Five-Year Review. The Environmental Law Reporter 27(10504).

Dernbach, J.C. 1998. Sustainable development as a framework for national governance. Case W. Res. Law Rev. 49(1):1-103.

Dernbach, J.C., ed. 2002. Stumbling toward Sustainability. Washington, DC: Environmental Law Institute.

Dernbach, J.C. 2003. Achieving sustainable development: The centrality and multiple facets of integrated decision making. Ind. J. Global Leg. Stud. 10(1):247-284.

Dernbach, J.C., ed. 2009. Agenda for a Sustainable America. Washington, DC: Environmental Law Institute.

DOE (U.S. Department of Energy). 2011. Better Buildings. U.S. Department of Energy, Energy Efficiency and Renewable Energy [online]. Available: http://www1.eere.energy.gov/buildings/betterbuildings/ [accessed Apr. 19, 2011].

EDF (Environmental Defense Fund). 2011. The Cap and Trade Success Story. Environmental Defense Fund, New York [online]. Available: http://www.edf.org/page.cfm?tagID=1085 [accessed Apr. 18, 2011].

Egan, T. 2006. The Worst Hard Time: The Untold Story of Those Who Survived the Great American Dust Bowl. New York: Houghton Mifflin.

ELI (Environmental Law Institute). 1991. Public Participation in Environmental Regulation. Washington, DC: Environmental Law Institute.

ELI. 1995. Rediscovering the National Environmental Policy Act: Back to the Future. Washington, DC: Environmental Law Institute.

EPA (U.S. Environmental Protection Agency). 2008. EPA's Report on the Environment. EPA/600/R-07/045F. U.S. Environmental Protection Agency, Washington, DC. May 2008 [online]. Available: http://www.epa.gov/ncea/roe/docs/roe_final/EPAROE_FINAL_2008.PDF [accessed Apr. 19, 2011].

EPA. 2010. Summary of Federal Insecticide, Fungicide, and Rodenticide Act (FIFRA). U.S. Environmental Protection Agency [online]. Available: http://www.epa.gov/agriculture/lfra.html#Summary%20of%20the%20Federal%20Insecticide,%20Fungicide,%20and%20Rodenticide%20Act [accessed June 9, 2011].

Feldman, I.R. 2009. Business and industry: Transitioning to sustainability. Pp. 71-92 in Agenda for a Sustainable America, J.C. Dernbach, ed. Washington, DC: Environmental Law Institute.

Fiksel, J., T. Graedel, A.D. Hecht, D. Rejeski, G.S. Sayler, P.M. Senge, D.L. Swackhamer, and T.L. Theis. 2009. EPA at 40: Bringing environmental protection into the 21st century. Environ. Sci. Technol. 43(23):8716-8720.

Fitzgerald, J. 2010. Emerald Cities: Urban Sustainability and Economic Development. New York: Oxford University Press.

Fox, S. 1981. The American Conservation Movement: John Muir and His Legacy. Madison, WI: University of Wisconsin Press.

Hawken, P. 2007. Blessed Unrest: How the Largest Movement in the World Came into Being and Why No One Saw It Coming. New York: Viking Press.

Hays, S.P. 1959. Conservation and the Gospel of Efficiency: The Progressive Conservation Movement, 1890-1920. Cambridge, MA: Harvard University Press.

Hays, S.P. 1987. Beauty, Health, and Permanence: Environmental Politics in the United States, 1955-1985 (Studies in Environment and History). Cambridge: Cambridge University Press.

IPCC (Intergovernmental Panel on Climate Change). 2007. Climate Change 2007: The Physical Science Basis. Cambridge: Cambridge University Press [online]. Available: http://www.ipcc.ch/publications_and_data/publications_ipcc_fourth_assessment_report_wg1_report_the_physical_science_basis.htm [accessed Apr. 18, 2011].

IUCN (International Union for Conservation of Nature). 1980. World Conservation Strategy. Living Resource Conservation for Sustainable Development. Gland, Switzerland: IUCN [online]. Available: http://data.iucn.org/dbtw-wpd/edocs/WCS-004.pdf [accessed Apr. 18, 2011].

Lazarus, R.J. 2004. The Making of Environmental Law. Chicago: University of Chicago Press.

Leopold, A. 1986. Game Management. Madison, WI: University of Wisconsin Press.

Marsh, G.P. 1864. Man and Nature; or, Physical Geography as Modified by Human Action. Cambridge, MA: Belknap Press of Harvard University Press.

Mazmanian, D.A., and M.E. Kraft. 2009. Toward Sustainable Communities: Transition and Transformation in Environmental Policy, 2nd Ed. Cambridge: MIT Press.

MEA (Millennium Ecosystem Assessment). 2005. Ecosystems and Human Well-Being: Synthesis. Washington, DC: Island Press.

Norse, E.A. 2005. Destructive fishing practices and evolution of the marine ecosystem-based management paradigm. Pp. 101-114 in Benthic Habitats and the Effects of Fishing, P.W. Barnes, and J.P. Thomas, eds. American Fisheries Society Symposium 41. Bethesda, MD: American Fisheries Society.

NRC (National Research Council). 1999. Our Common Journey: A Transition toward Sustainability. Washington, DC: National Academy Press.

NRC. 2010a. Advancing the Science of Climate Change. Washington, DC: The National Academies Press.

NRC. 2010b. Limiting the Magnitude of Future Climate Change. Washington, DC: The National Academies Press.

OECD (Organisation for Economic Co-operation and Development). 2001. OECD Environmental Strategy for the First Decade of the 21st Century. Organisation for Economic Co-operation and Development, May 21, 2001 [online]. Available: http://www.oecd.org/dataoecd/33/40/1863539.pdf [accessed Apr. 18, 2011].

OECD. 2005. OECD Environmental Performance Reviews: United States. Paris: OECD.

OECD. 2009. Declaration on Green Growth. C/MIN(2009)5/ADD1/FINAL. Organisation for Economic Co-operation and Development, June 25, 2009 [online]. Available: http://www.greengrowth.org/download/2009/news/OECD.declaration.on.GG.pdf [accessed Apr. 18, 2011].

OECD. 2011. Towards Green Growth: Green Growth Strategy Synthesis Report. C/MIN(2011)4. Organisation for Economic Co-operation and Development, May 2, 2011 [online]. Available: http://www.oecd.org/officialdocuments/publicdisplaydocumentpdf/?cote=C/MIN(2011)4&docLanguage=En [accessed June 7, 2011].

OMB (Office of Management and Budget). 2003. Informing Regulatory Decisions: 2003 Report to Congress on the Costs and Benefits of Federal Regulations and Unfunded Mandates on State, Local, and Tribal Entities. Office of Management and Budget, Washington, DC [online]. Available: http://www.whitehouse.gov/sites/default/files/omb/assets/omb/inforeg/2003_cost-ben_final_rpt.pdf [accessed Apr. 18, 2011].

O'Riordan, T., and J. Cameron, eds. 1994. Interpreting the Precautionary Principle. London: Earthscan.

PACEO (President's Advisory Council on Executive Organization). 1970. Ash Council Memo: Federal Organization for Environmental Protection, Memorandum for the President, from President's Advisory Council on Executive Organization, Executive Office of the President, Washington, DC. April 29, 1970 [online]. Available: http://www.epa.gov/history/org/origins/ash.htm [accessed Apr. 18, 2011].

PCSD (President's Council on Sustainable Development). 1996. Sustainable America: A New Consensus for Prosperity, Opportunity, and a Healthy Environment for the Future. President's Council on Sustainable Development, Washington, DC [online]. Available: http://clinton2.nara.gov/PCSD/Publications/TF_Reports/amer-top.html [accessed Apr. 18, 2011].

Pinchot, G. 1910. The Fight for Conservation. New York: Doubleday, Page & Company.

Pinchot, G. 1947. Breaking New Ground. New York: Harcourt Brace.

Porter, M.E., and M.R. Kramer. 2011. The Big Idea: Creating Shared Value. Harvard Business Review, Jan.-Feb. 2011[online]. Available: http://hbr.org/2011/01/the-big-idea-creating-shared-value/ar/1 [accessed Apr. 18, 2011].

Reiger, J.F. 1975. American Sportsmen and the Origins of Conservation. New York: Winchester Press.

Richardson, B.J., and S. Wood. 2006. Environmental Law for Sustainability. Portland, OR: Hart Publishing.

Rodgers, W.H., Jr. 1994. Environmental Law, 2nd Ed. St. Paul, MN: West Publishing Co.

Rodgers, W.H., Jr. 1999. Environmental Law, 2nd Ed. St. Paul, MN: West Publishing Co.

Roosevelt, T. 1907. Seventh Annual Message to the Senate and House of Representatives, December 3, 1907 [online]. Available: http://www.presidency.ucsb.edu/ws/index.php?pid=29548#axzz1JuA6cqEy [accessed Apr. 18, 2011].

Ruffing, K.G. 2010. The role of the Organization for Economic Cooperation and Development in environmental policy making. Rev. Environ. Econ. Policy 4(2):199-220.

Sarkar, R. 2009. P. xvi in International Development Law: Rule of Law, Human Rights, and Global Finance. Oxford: Oxford University Press.

Schlesinger, S.C. 2003. Act of Creation: The Founding of the United Nations. Boulder, CO: Westview Press.

Sen, A. 1999. Development as Freedom. New York: Random House.

TEEB (The Economics of Ecosystems and Biodiversity). 2010. Mainstreaming the Economics of Nature: A Synthesis of the Approach, Conclusions and Recommendations of TEEB. Malta: Progress Press [online]. Available: http://www.teebweb.org/LinkClick.aspx?fileticket=bYhDohL_TuM%3d&tabid=1278&mid=2357 [accessed Apr. 19, 2011].

UN (United Nations). 1997. Programme for the Further Implementation of Agenda 21. A/RES/S-19/2. United Nations General Assembly, September 19, 1997 [online]. Available: http://www.un.org/documents/ga/res/spec/aress19-2.htm [accessed May 5, 2011].

UN. 2000. Millennium Declaration. United Nations General Assembly, September 8, 2000 [online]. Available: http://www.un.org/millennium/declaration/ares552e.htm [accessed May 5, 2011].

UN. 2002. Johannesburg Plan of Implementation. World Summit on Sustainable Development, 26 August-4 September 2000, Johannesburg, South Africa United Nations Conference on Trade and Development. New York: United Nations [online]. Available: http://www.unctad.org/en/docs/aconf199d20&c1_en.pdf [accessed May 5, 2011]

UNCED (United Nations Conference on Environment and Development). 1992a. Agenda 21. United Nations Conference on Environment and Development, Rio de Janeiro, Brazil, 3-14 June 1992 [online]. Available: http://www.cuttingthroughthematrix.com/articles/Agenda21.pdf [accessed Apr. 18, 2011].

UNCED. 1992b. Report of the United Nations Conference on Environment and Development, Rio de Janeiro, 3-14 June 1992, Annex I. Rio Declaration on Environment and Development. A/CONF.151/26 (Vol. I). United Nations General Assembly [online]. Available: http://www.un.org/documents/ga/conf151/aconf15126-1annex1.htm [accessed Apr. 15, 2011].

UNCSD (United Nations Conference on Sustainable Development). 2011. Earth Summit 2012: Objectives and Themes [online]. Available: http://www.earthsummit2012.org/index.php/objectives-and-themes [accessed Apr. 19, 2011].

UNDP (United Nations Development Programme). 1999. Human Development Report 1999. New York: Oxford University Press [online]. Available: http://hdr.undp.org/en/media/HDR_1999_EN.pdf [accessed June 9, 2011].

UNEP (United Nations Environment Programme). 2011a. Environment for Development [online]. Available: http://www.unep.org/environmentalgovernance/ [accessed Apr. 19, 2011].

UNEP. 2011b. Towards a Green Economy: Pathways to Sustainable Development and Poverty Eradication. A Synthesis for Policy Makers. United Nations Environment Programme. [online]. Available: http://www.unep.org/greeneconomy/Portals/88/documents/ger/GER_synthesis_en.pdf [accessed Apr. 19, 2011].

UNESA (United Nations Department of Economic and Social Affairs). 2011. National Sustainable Development Strategies: The Global Picture 2010. U. N. Department of Economic and Social Affairs, Sustainable Development Division [online]. Available: http://www.un.org/esa/dsd/dsd_aofw_nsds/nsds_pdfs/NSDS_map.pdf [accessed Apr. 19, 2011].

UNGA (United Nations General Assembly). 1983. Process of Preparation of the Environmental Perspective to the Year 2000 and Beyond. Resolution 38/161. United Nations General Assembly, December 19, 1983 [online]. Available: http://www.un.org/documents/ga/res/38/a38r161.htm [accessed June 8, 2011].

UNGA. 2010. Implementation of Agenda 21, the Programme for the Further Implementation of Agenda 21 and the outcomes of the World Summit on Sustainable Development. Resolution 64/236. United Nations General Assembly, March 31, 2010 [online]. Available: http://www.uncsd2012.org/files/OD/ARES64236E.pdf [accessed June 8, 2011].

Van Hise, C.R. 1927. The Conservation of Natural Resources in the United States. New York: Macmillan.

WBCSD (World Business Council for Sustainable Development). 2011. Vision 2050. World Business Council for Sustainable Development [online]. Available: http://www.wbcsd.org/templates/TemplateWBCSD5/layout.asp?MenuID=1 [accessed May 31, 2011].

WCED (United Nations World Commission on Environment and Development). 1987. Our Common Future. Oxford: Oxford University Press.

Weiss, J.D. 2009. Local governance and sustainability: Growing progress, major challenges. Pp. 43-56 in Agenda for a Sustainable America, J.C. Dernbach, ed. Washington, DC: Environmental Law Institute.

3

A Sustainability Framework for EPA

INTRODUCTION

EPA asked the Committee on Incorporating Sustainability in the U.S. EPA to address the question "What should be the *operational framework* for sustainability for EPA?" The primary design feature of such a framework is that it support and guide EPA's actions to further sustainability goals. On the basis of presentations made to the committee by various representatives of EPA and other experts, the sustainability literature, and the experience of committee members, certain other key attributes of the operational framework are suggested. The framework will more likely be successful if it is (1) transparent and clear, (2) practical to implement within the existing program structure of EPA, (3) leads to goals and objectives that can be measured and reported publicly, (4) provides flexibility to deal with scientific, technical and economic developments over long time frames (more than 5 years), (5) works consistently with the current risk assessment/risk management paradigm, and (6) facilitates decision making that supports EPA's ongoing mission to protect human health and the environment.

EPA has made progress in implementing sustainability within the agency and in providing opportunities to advance sustainable practices by other agencies and organizations in the United States through its programs, research and development (R&D), and regulatory mandates. However, clarifying its intent to incorporate sustainability concepts and practices across and within the organization will help to accelerate progress toward achieving EPA's overarching sustainability goals as discussed later in the chapter. Although media-specific approaches still exist within EPA, one strength of a sustainability approach is that it encourages cross-media approaches. Additionally, the committee does not anticipate EPA will use the Sustainability Framework for all decisions, but does anticipate that over

35

time there will be an increasing use of the framework. As with all decision making at EPA, uncertainty needs to be acknowledged and addressed, the approach needs to be transparent, and key stakeholders need to be engaged.

Figures 3-1 and 3-2 illustrate the committee's recommended Sustainability Framework for EPA. The overall approach is driven by sustainability principles and goals and involves setting, meeting, and reporting on measurable performance objectives. As such, the approach reflects an overall management system framework for sustainability. The framework includes a specific "Sustainability Management and Assessment" component for incorporating sustainability into individual EPA decisions and actions, represented by the inset in Figures 3-1 and 3-2. The Sustainability Assessment and Management process is intended to be applicable to all types of issues, including human health and ecological risks. Similar approaches have been used successfully in both the private and public sectors, including several examples described in Box 3-1. In addition, the committee was informed by several efforts to synthesize the literature on sustainable development and propose sustainability frameworks (including Graedel and Klee 2002; Marshall and Toffel 2005; Porritt 2007; Jabareen 2008). In fact, EPA has been a user and promoter of environmental-management-system frameworks. This topic was the subject of Executive Order 13148, which was later reaffirmed by President Bush's administration in a 2006 memorandum "Commitment to the Integration and Utilization of Environmental Management Systems" (EPA 2006). The agency has also prepared guidance for developing environmental-management-system frameworks for organizations and businesses, noting in its introductory remarks to these entities, "As one of your organization's leaders, you probably know that interest in environmental protection and sustainable development is growing each year. You might hear about these issues from customers, the public, or others. Like many, your organization may be increasingly challenged to demonstrate its commitment to the environment. Implementing an [environmental management system] can help you meet this challenge in several important ways" (EPA 2008).

THE SUSTAINABILITY FRAMEWORK

A management system framework will accelerate incorporation of sustainability into the operational activities of EPA, which has many motivated and committed professionals who enjoy their jobs and do them well with the goal of protecting public health and ecosystem health. The framework will guide their management of competing priorities and the "pushes and pulls" inherent in their roles by providing a basis for setting priorities based in part on sustainability considerations. In particular, decision makers in the agency at all levels have a special responsibility in considering and making the trade-offs and finding balances inherent in a sustainability framework. This recommended operational framework for sustainability will accelerate alignment and culture

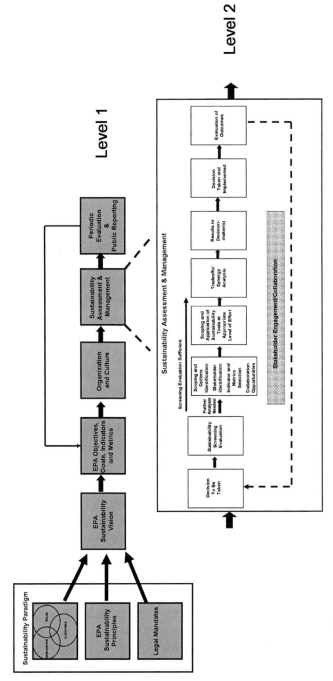

FIGURE 3-1 A framework for EPA sustainability decisions

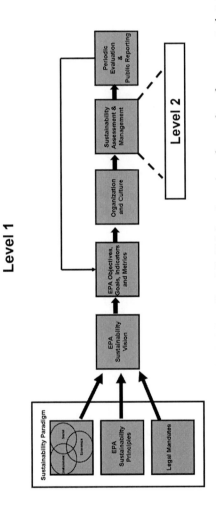

FIGURE 3-2 A framework for EPA sustainability decisions (Level 1). This level consists of a number of components that define the agency-wide process.

BOX 3-1
Examples of Management System Frameworks for Sustainability

A management-system approach, built on vision, objectives, goals, and metrics, is a commonly used approach to changing cultures and delivering significant course corrections and performance improvement. A well-known example of this approach is the setting of the U.S. aspiration to send a man to the moon, with a safe return, by President John F. Kennedy. This vision and goal set into motion strategic planning and goal setting that helped the country to achieve what was then, and is now, a remarkable breakthrough. Many other government agencies have used this kind of approach—for example, in the eradication of certain infectious diseases and reduction of smoking rates. Currently, the U.S. Geological Survey (USGS) is using a similar approach (USGS 2007, 2010) by realigning its management and budget structure to support a newly developed science strategy designed to address the major scientific issues facing the nation, including global climate change, water resources, natural hazards, energy and minerals, ecosystems, and data integration (USGS 2007, 2010).

Several NGOs have been involved in collaborative sustainability activities, including the World Resources Institute, the Environmental Defense Fund, and the World Wildlife Fund. Additionally, sustainability has been embraced as a cost-effective organizing principle by the private industry. For example, Proctor & Gamble (P&G) recently released a long-term vision in sustainability and new 2020 goals related to products, operations, and social responsibility (P&G 2011). Similarly, Unilever released its Sustainable Living Plan which includes 50 goals related to the company's environmental impacts of its products (Unilever 2011). Wal-Mart has also made enormous changes in energy and resource consumption in their supply chains (Quinn 2009).

---

change in the agency and provide a platform for more integrated decision making in the future.

EPA is already engaged in many projects and approaches that further sustainability aims, but adoption of this approach *with staged and programmatic implementation* will lead to a growing body of agency successes and experiences with the incorporation of sustainability. The adoption of the overall management system approach could occur quickly, and the use of the Sustainability Assessment and Management process can be phased in over time on agency decisions and actions.

Much of the acceleration of progress will occur as the culture of EPA moves toward incorporation of sustainability, and as EPA's intentions and goals in sustainability become clear to employees. The good work of EPA already in progress is encouraged, and more experimentation in the program offices and in the regions will help lead the way to a new EPA culture.

The Sustainability Framework is organized into a two-level process and is presented in Figure 3-1. Level 1 consists of a number of components that define the agency-wide process (Figure 3-2). Level 1 components are depicted in gray and described in this chapter. Level 2 (Figure 4-1) articulates the elements of the Sustainability Assessment and Management component that are described in Chapter 4.

THE SUSTAINABILITY FRAMEWORK: LEVEL 1 COMPONENTS

Level 1 of the Sustainability Framework, described below, includes the sustainability paradigm, principles, and legal mandates that feed into the process; EPA's sustainability vision as well as objectives, goals, and indicators and "metrics" (see Appendix C); organization and culture; sustainability assessment and management; and periodic evaluation and public reporting activities.

Sustainability Paradigm, EPA Sustainability Principles, and Legal Mandates

EPA needs to specify and acknowledge a set of principles and assumptions that underlie its approach to sustainability. First, the committee recommends that EPA formally adopt as its sustainability paradigm the "Three Pillars" approach of "Social," "Environment," and "Economic" dimensions of sustainability as a well-recognized and established model for evaluating sustainability in its decisions. The committee recognizes that there are other sustainability models and paradigms, but the "Three Pillars" model has stood the test of time in many organizational implementations owing to its utility and its relative simplicity and clarity. For example, the United Nations et al. (2003) has noted that this three-pillar approach to sustainable development is widely held, wide-ranging, and complex. EPA already has a strong focus on the environmental component of the paradigm and on the social pillar as it relates to managing human health risks. However, an integrated approach across all dimensions has not been the usual practice of EPA. The "Three Pillars" approach should become a key part of educating all employees about agency philosophy, principles, and performance expectations. It is a powerful yet simple paradigm for explaining and incorporating the key dimensions of sustainable development. (The "Three Pillars" approach is also commonly known as the "Triple Bottom Line" approach. As explained in Chapter 2, the roots of both are the same.) Neither the pillar labeled "environment" nor the "social" pillar explicitly convey the inclusion of human health and well-being. The social pillar traditionally includes human health aspects (Chapter 2) and the inclusion of human health under it would be made transparent by defining it as the "social pillar (including human health)." This inclusion would clarify the important place of human health in the Sustainability Framework. The committee recognizes that the social pillar also includes environmental justice and employment, among other things, but human health deserves explicit mention because of its historical role as part of the agency's mission.

It is important for EPA to optimize all three of the pillars of sustainability. Although addressing economic issues is not a core part of EPA's mission, it is explicitly part of the definition of sustainability (NEPA[1969], E. O.13514[2009]). Inherent in the definition of sustainability is the recognition of the importance of the three pillars. Optimizing this three pillar approach is key to the Trade-off Analysis and Synergy component of the Sustainability Framework. The economic factors in question will depend, in part, on the decision at hand, and how to address those factors will depend on the tool EPA chooses in its analysis. Decisions that further one of the three pillars should, to the extent possible, further the other two. Where EPA has the legal authority to consider economic factors, integrating sustainability into EPA's decision making means furthering all three pillars as much as possible at the same time.

Second, and equally important, EPA could benefit from formally developing, adopting, and publishing a set of broad "EPA Sustainability Principles," which underlie all agency policies and programs. These principles would guide the agency's implementation of regulatory mandates and discretionary programs in ways to optimize benefits as they relate to the three pillars—social, environmental, and economic. Several key principles of public administration include openness and transparency, reliability, accountability, efficiency, and effectiveness (OECD 1999). A wide variety of principles have been articulated by government agencies, nongovernmental organizations (NGOs), international governmental organizations (IGOs), and companies. Box 3-2 identifies sample principles that have been used internationally or by specific countries. The committee included these principles for illustration purposes only. Although EPA has had agency-wide guiding principles (EPA 1997) and may utilize or research sustainability principles for particular applications (EPA 2007), it has yet to clearly articulate a broad set of sustainability principles to guide decisions agency-wide. The most widely cited and used set of principles in the world are the sustainable development principles from the Rio Declaration of 1992 (Chapter 2). Others have been developed within the OECD, where the United States has joined in the consensus leading to their adoption. Some of the key dimensions of the principles that EPA should debate and adopt include intergenerational and intragenerational equity,[1] justice, and a holistic-systems approach to environmental problems and solutions. Abbott and Marchant (2010) stated that a "notable aspect of sustainability

---

[1] Professor Edith Brown Weiss explained intergenerational equity as having three elements, including an intragenerational aspect. First, each generation should conserve the options of future generations by conserving "the diversity of the natural and cultural resource base." Second, each generation is entitled to a quality of planet enjoyed by prior generations, and also has an obligation to pass to the next generation a quality of planet that is no worse than it received. Third, all people in the current generation should have the same minimal level of access to this legacy. Because poverty and environmental degradation are inseparably linked, equity within the current generation is necessary for equity between generations (Weiss 1989).

BOX 3-2
Selected International and National Sustainability Principles

Some selected principles developed internationally to guide sustainability efforts
are listed below.

Rio Declaration (described in Chapter 2)
- "In order to achieve sustainable development, environmental protection shall
  constitute an integral part of the development process and cannot be considered in isolation from it."
- Precautionary approach: "Where there are threats of serious or irreversible
  damage, lack of full scientific certainty shall not be used as a reason for postponing cost-effective measures to prevent environmental degradation."
- Intergenerational equity: The Rio Declaration acknowledges the need "to
  equitably meet developmental and environmental needs of present and future
  generations."
- Internalization of environmental costs: The "approach that the polluter should,
  in principle, bear the cost of pollution," is reflected in varying degrees throughout U.S. environmental law.
- "Environmental issues are best handled with participation of all concerned
  citizens, at the relevant level" (UNCED 1992).

OECD Environmental Strategy for the First Decade of the Twenty-first Century
- "Regeneration: Renewable resources shall be used efficiently and their use
  shall not be permitted to exceed their long-term rates of natural regeneration."
- "Substitutability: Non-renewable resources shall be used efficiently and their
  use limited to levels which can be offset by substitution by renewable resources
  or other forms of capital."
- "Assimilation: Releases of hazardous or polluting substances to the environment shall not exceed its assimilative capacity; concentrations shall be kept
  below established critical levels necessary for the protection of human health
  and the environment."
- "Avoiding Irreversibility: Irreversible adverse effects of human activities on
  ecosystems and on biogeochemical and hydrological cycles shall be avoided"
  (OECD 2001).

is its holistic and cross-cutting nature—it cannot be achieved by any single rule,
statute, or agency" (p.1924).

Third, EPA would benefit from making sustainability part of the "how" EPA
goes about implementing its regulatory authorities and objectives, thus creating
more value in its work but not at the expense of its responsibilities under the law.
Implementing its regulatory mandates is the core work of the agency. Also, under
the Sustainability Framework, EPA would continue to promote human well-being
and protect human health in implementing core mandates while more deliberately
addressing the other dimensions (e.g., giving ecosystem well-being higher prior-

Australia's National Strategy for Ecologically Sustainable Development's Guiding Principles
- "Decision making processes should effectively integrate both long and short-term economic, environmental, social and equity considerations."
- "Where there are threats of serious or irreversible environmental damage, lack of full scientific certainty should not be used as a reason for postponing measures to prevent environmental degradation."
- "The global dimension of environmental impacts of actions and policies should be recognised and considered."
- "The need to develop a strong, growing and diversified economy which can enhance the capacity for environmental protection should be recognized."
- "The need to maintain and enhance international competitiveness in an environmentally sound manner should be recognized."
- "Cost effective and flexible policy instruments should be adopted, such as improved valuation, pricing and incentive mechanisms."
- "Decisions and actions should provide for broad community involvement on issues which affect them" (AESDSC 1992).

Canada's Federal Sustainable Development Strategy
The Federal Sustainable Development Act (FSDA) states that "The Government of Canada accepts the basic principle that sustainable development is based on an ecologically efficient use of natural, social and economic resources." The government of Canada's approach to sustainable development therefore reflects a commitment to minimize the environmental impacts of its policies and operations as well as maximize the efficient use of natural resources and other goods and services. Canada's environmental policy is guided by the precautionary principle and is reflected in the Federal Sustainable Development Strategy (FSDS) as required by the Federal Sustainable Development Act, which states that the Minister of Environment must "develop a Federal Sustainable Development Strategy based on the precautionary principle" (Environment Canada 2010).

ity than in previous EPA efforts). EPA is in a unique position to further sustainability opportunities for all stakeholders by implementing this framework and by working collaboratively to find more optimal sustainability solutions.

EPA Sustainability Vision

A clear statement of sustainability vision is an important step to helping all EPA employees understand and execute their responsibilities in sustainability. EPA needs to formally develop and specify its vision for sustainability. Vision, in the sense discussed here, is a future state that EPA is trying to reach or is trying to help the country or the world to reach. There is literature evaluating whether visions promote organizational effectiveness (Zaccaro and Banks 2001, Kantabutra 2010, O'Connell et al. 2011), including conceptual models developed to evaluate how visions promote organizational effectiveness suggest that they do influence organizational effectiveness in several ways (Projasek 2003, Parrish 2010), including by "providing a source of envisioned empowerment that motivates followers" (Zaccaro and Klimoski 2001). Although EPA has always had clear statements of mission, times call for a very clear articulation of the sustainability vision that EPA is trying to help deliver through its work. It is important that EPA debate and adopt a sustainability vision as an organic process to guide the agency forward.

EPA's stated mission—to protect human health and the environment—appropriately reflects the agency's statutory authority. In addition to its mission, EPA lists seven specific purposes that explain the mission (see Chapter 1 for discussion about EPA's mission). For example, one purpose for EPA is "to ensure that environmental protection is an integral consideration in U.S. policies concerning natural resources, human health, economic growth, energy, transportation, agriculture, industry, and international trade, and these factors are similarly considered in establishing environmental policy" (EPA 2011). Another purpose is "to ensure that environmental protection contributes to making our communities and ecosystems diverse, sustainable and economically productive" (EPA 2011). Together with the mission statement, those purposes address the question; Why does EPA exist? They also reflect sustainability principles. Still, EPA may wish to reconsider these statements at the same time it finalizes its sustainability vision to assure coherence and clarity.

EPA Objectives, Goals, Indicators, and Metrics

Once articulated, the sustainability vision will be a touchstone in the future that guides the setting of objectives and goals. EPA could benefit from considering a goal-setting approach that starts with a long-term vision. One such approach, for illustration, is a "backcasting" approach developed for sustainability (Holmberg and Robert 2000, Natural Step 2011).

In backcasting, a future vision is articulated. With that vision in mind, goals and objectives are developed for immediate implementation that in a shorter term will help the organization to make significant progress toward the long-term vision (Natural Step 2011). Setting several breakthrough 3-5 year objectives[2] could assist EPA in delivering a new level of performance in driving more sus-

---

[2] Commonly referred to in the business community, breakthrough objectives are goals that extend far beyond the current capabilities and experiences of an organization and require new strategies and approaches to ensure successful attainment of these goals. These objectives are generally designed to improve performance throughout an organization.

tainable outcomes and most important, will guide EPA employees in seeing the commitment of agency leadership to new ways of operating. For example, EPA can specify several 3-5 year objectives and associated short-term measureable goals that advance the agency and the country toward the sustainability vision. For illustration purposes only, EPA could set a 3-5 year objective that "by 2015, the EPA will have facilitated 25 green infrastructure projects that deliver regulatory performance comparable to conventional pollution control systems, but at lower costs, with higher conservation values and outcomes, and measurable net job creation."

This kind of objective would be a strong signal of the need to shift to place-based sustainability projects and experimentation with green infrastructure approaches. The indicators for assessing performance against this objective could be the number of green infrastructure projects completed, associated cumulative value of cost savings and ecosystem service value gained, and human health risk reduction. Again, this example is only for illustrative purposes; however, the pursuit of breakthrough 3-5 year objectives, chosen well, could help EPA reshape its culture and its cross-program teamwork and help all employees—headquarters and regional—align their professional goals and activities. The NRC (2006) noted that "successfully targeting and sustaining programs linking knowledge with action for sustainability generally requires a clear, readily understood statement of the beneficial outcomes that successful project completion would deliver. Operationally, this translates into the articulation of clear, broadly shared goals, and the development and operational measurement of generally accepted indicators of goal achievement" (p.18).

Organization and Culture

Identifying and communicating the objectives selected, the indicators and the associated metrics needed for implementation is the next step. EPA could benefit from staying focused on implementation and minimizing structural changes in the organization to help it embrace the new goals and objectives in the current organizational configuration. To a great extent, the committee's primary focus is much more on developing a new culture of sustainability in the agency (Chapter 6) rather than on organizational changes. The agency could benefit from specifically focusing on education, learning, and alignment of senior staff, and eventually all employees, with the new sustainability approach and the roles everyone can play in helping to accelerate progress. The fastest progress will be made through empowering the employees of EPA and by building a strong sustainability culture at the agency.

The Office of Research and Development (ORD) has already begun to organize its programs around sustainability-related themes (Figure 3-3), which is consistent with the "sustainability journey" described, and several EPA regional programs have also begun to think about their programs and projects in a new

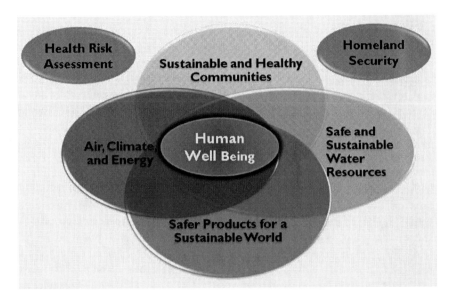

FIGURE 3-3 Reorganization themes of the Office of Research and Development.
SOURCE: Hecht 2010.

sustainability-oriented manner (Chapter 7). Organizing programs and themes around accomplishing sustainability objectives can accelerate progress.

Sustainability Assessment and Management

The overall framework covers the highest levels of visioning and goal setting for EPA, but it also covers the daily implementation of sustainability as part of the ongoing decisions and actions. The component of the framework "Sustainability Assessment and Management," provides a decision-making approach for incorporation of sustainability into the daily work of the agency. It is intended to deal with the "nuts and bolts" of implementation. Chapter 4 of this report is devoted to describing the elements of the Sustainability Assessment and Management component and the analytic tools for use in applying it.

EPA at all levels, offices, and regions faces daily decisions and potential actions in various contexts—for example, routine permitting, enforcement actions, reviewing significant new-use chemical applications, development of regulations, and major policy initiatives. The general approach recommended by the committee for sustainability assessment and management—laid out as Level 2 in Figure 3-3 and discussed in greater depth in Chapter 4—will need to be shaped for application in EPA's various programs, and some routine decision making may not be best served by formal sustainability reviews for each project.

EPA will need to decide what kinds of activities or actions to address in a sustainability assessment process. Adopting a sustainability framework is discretionary, and EPA will choose where to focus its attention and resources. For example, the agency may wish to focus on major new rules, programs, or policies; priorities for program improvement; and complex and important emerging issues, such as the periodic review of the impacts of biofuels production and use required by Congress (Public Law 110-140). Another option could be for EPA's program and regional offices to be tasked with selecting priority initiatives or activities in their fields of work for sustainability assessments. The choices made may or may not be related to the goals set at the beginning. Certain categories of actions, such as routine permits, grants, and enforcement actions, could be excluded altogether from consideration for sustainability review, or the agency could decide to infuse sustainability thinking and practices into routine decision-making processes as they are periodically reviewed and updated.

Over time, EPA will need to develop approaches to decide when and how to incorporate sustainability considerations into agency decisions and actions, particularly those significant enough to be moved through the Sustainability Assessment and Management approach (Level 2); staff training on this approach will be needed. The scale and depth of the analysis and other aspects of the process, including stakeholder involvement, will need to be commensurate with the decision type and potential sustainability impact. The committee realizes that it will take some time to effect this change in culture and process.

Periodic Evaluation and Public Reporting

The collective actions of an EPA program and the agency as a whole (and its partners in communities and other agencies) will have an impact on the sustainability indicators (linked to breakthrough objectives and goals) of the agency as well as on many other indicators. The metrics for these indicators should be evaluated annually, checking for progress over time and informing needs for change within the agency in terms of organization, focus, and resource allocation.

For purposes of this report, the committee will refer to "metrics" as the measured values used to assess specific indicators of progress (see Box 3-3 and additional discussion of indicators and metrics in Chapter 4). Indicators, in general, are measures of impact that can be aggregated to track overall progress. EPA would benefit from an effort to develop metrics (measurement systems) for those indicators that could serve as a "dashboard of progress" for the agency and its administration. The selection of a slate of metrics is a challenging endeavor, requiring appropriate resources. It is necessary to directly link selected metrics to EPA's vision as well as to the long-term goals and objectives of the agency. Once objectives and indicators are identified and metrics specified, they can become part of the long-term commitments of the agency to be reported annually in public reports and administrator summaries of accomplishments.

BOX 3-3
Goal, Indicator, and Metric

Goal—what is specifically sought to be achieved. The goal is determined through the use of measured indicators.
Example: Reducing mercury emissions from electric utility steam generating units.

Indicator—"A summary measure that provides information on the state of, or change in, a system" (OECD 2011b), that is, what is being measured.
Example: Mass of mercury emitted per heat energy input, for example, pounds per gigawatt hours.

Metric—"Defines the unit of measurement or how the indicator is being measured" (OECD 2011a).
Example: Using the first definition, an example metric would be [grams Hg (of mercury)/Kwh (of energy input)]. The description below illustrates the metric in the latter sense—specifying exactly how one arrives at the measure—so that any two individuals in different institutions would come up with the same number.

Example: A mercury continuous-emissions monitoring system is used to measure molecules of the element Hg (mercury) in the stack gas as they pass through a sorbent trap. This measure is reduced to an hourly mass emissions rate. Then the indicator is calculated from a formula that uses the Hg hourly concentration, the flow rate of the stack gas, the electrical load, the diluent gas concentration, and moisture data. This indicator is then compared with a reference value to determine whether the unit is in compliance (EPA 2010)

A set of indicators and associated metrics (associated with goals and objectives) and indicators associated with international reporting protocols could be published periodically by EPA to inform its employees and the public about progress toward national objectives and goals. Over time, this reporting will become a key part of a new culture at EPA that is linked to delivering measureable progress toward sustainability. Goals, objectives, and indicators and associated metrics would benefit from being periodically reviewed for relevance to improving sustainability and for EPA's ability to have an impact on them positively through their actions.

Federal agencies are required under the Government Performance and Results Act (GPRA) to prepare strategic plans, annual performance plans, and annual performance reports (31 U.S.C. § 1115). Agencies must clearly identify high-priority performance goals in their strategic plans, annual performance plans, and annual reports and are expected to internally review performance related to their goals. The Office of Management and Budget (OMB) directs senior agency leaders to hold "goal-focused, data-driven reviews at least once every quarter to review progress on agency priorities and to assure that follow-up steps are taken to increase the likelihood of achieving better outcomes, higher productivity, and priority goals" (OMB 2010). Meeting GPRA requirements and developing sustainability indicators and associated metrics could be coupled, for example, under EPA's 5 year Strategic Plan (EPA 2010).

FINDINGS AND RECOMMENDATIONS

3.1. Key Finding: Organizations that most effectively integrate sustainability into their work and culture are those that base their programs on clear principles, vision, strategic goals, and implementation processes (p.45-47).

3.1. Key Recommendation: EPA should adopt or adapt the comprehensive Sustainability Framework proposed in Figure 3-1. The proposed Sustainability Framework requires a comprehensive approach including specific processes for incorporating sustainability into decisions and actions. As part of the framework, EPA should incorporate upfront, consideration of sustainability options and analyses that cover the three sustainability pillars, as well as trade-off consideration into its decision making. This framework was developed with the intent that EPA could apply it to any decision to which a need arose.

3.2. Key Finding: Organizational incorporation of sustainability requires clear, broadly shared goals, and the development and operational measurement of generally accepted indicators and associated metrics of goal achievement (p.47).

3.2. Key Recommendation: For programs, EPA should set several strategic 3-5 year breakthrough objectives related to its sustainability implementation and its performance indicators and associated metrics. These goals would be designed to improve performance throughout the agency by extending beyond its current capabilities and experiences and requiring new strategies and approaches to ensure their attainment. EPA should begin periodic public sustainability reporting to transparently review its progress versus goals.

3.3. Key Finding: For successful implementation of sustainable-development principles in organizations and their work, the "Three Pillars" paradigm has a proven track record of effectiveness in the United States and globally. Practically all of EPA's activities under its environmental statutes are intended to protect human health and well-being. However, the pillar for "environment" may not convey the inclusion of human health and well-being under that heading. The committee proposes that EPA consider defining the social pillar as the "social pillar (including human health)" to clarify the important place of public health in the Sustainability Framework (p.40).

3.3. Key Recommendation: The committee recommends expressly including the term "health" in the social pillar to help ensure that EPA regulatory and scientific staff primarily concerned with human-health issues recognize their existing role in sustainability and recommends that EPA pay particular attention to explaining the role of human health in the social pillar, thereby ensuring that staff and stakeholders involved in the area of human health recognize that their activities are an integral part of EPA's sustainability work. Further, expressly including health in the social pillar will more clearly communicate outside of EPA the agency's role in that pillar of sustainability.

REFERENCES

Abbott, K.W., and G.E. Marchant. 2010. Institutionalizing sustainability across the federal government. Sustainability 2(7):1924-1942.

AESDSC (Australian Ecologically Sustainable Development Steering Committee). 1992. National Strategy for Ecologically Sustainable Development. Ecologically Sustainable Development, Australian Government, Canberra [online]. Available: http://www.environment.gov.au/about/esd/publications/strategy/index.html [accessed Apr. 19, 2011].

Environment Canada. 2010. Planning for a Sustainable Future: A Federal Sustainable Development Strategy for Canada. Sustainable Development Office, Environment Canada. October 2010 [online]. Available: http://www.ec.gc.ca/dd-sd/F93CD795-0035-4DAF-86D1-53099BD303F9/FSDS_v4_EN.pdf [accessed Apr. 19, 2011].

EPA (U.S. Environmental Protection Agency). 1997. EPA Strategic Plan. EPA/190-R-97-002. Office of the Chief Financial Officer, U. S Environmental Protection Agency, Washington, DC. September 1997 [online]. Available: https://courses.worldcampus.psu.edu/public/buried_assets/files/1-08-EPA%20Strategic%20Plan.pdf [accessed Apr. 19, 2011].

EPA. 2006. Commitment to the Integration and Utilization of Environmental Management Systems. Memorandum to Assistant Administrators, General Counsel, Inspector General, Chief Financial Officer, Associate Administrators, and Regional Administrators, from Stephen L. Johnson, the Administrator, U.S. Environmental Protection Agency, Washington, DC, October 16, 2006 [online]. Available: http://www.epa.gov/ems/docs/Final_EMS_Statement10.06.pdf [accessed Apr. 19, 2011].

EPA. 2007. Sustainability Research Strategy. EPA 600/S-07/001. Office of Research and Development, U.S. Environmental Protection Agency, Washington, DC. October 2007 [online]. Available: http://www.epa.gov/sustainability/pdfs/EPA-12057_SRS_R4-1.pdf [accessed Apr. 19, 2011].

EPA. 2008. Why Your Organization Should Have an EMS? Environmental Management Systems, U.S. Environmental Protection Agency, Washington, DC [online]. Available: http://www.epa.gov/ems/info/sme1.htm [accessed Apr. 19, 2011].

EPA. 2010. Fiscal Year 2011-2015 EPA Strategic Vision: Achieving Our Vision. U.S. Environmental Protection Agency, Washington, DC [online]. Available: http://www.epa.gov/planandbudget/strategicplan.html [accessed Apr. 18, 2011].

EPA. 2011. About EPA. U.S. Environmental Protection Agency [online]. Available: http://www.epa.gov/aboutepa/whatwedo.html [accessed Apr. 19, 2011].

Graedel, T.E., and R.J. Klee. 2002. Getting serious about sustainability. Environ. Sci. Technol. 36(4):523-529.

Hecht, A. 2010. U.S. Environmental Protection Agency. Presentation at the First Meeting on Incorporating Sustainability at the US EPA, December 14-15, 2010, Washington, DC.

Holmberg, J., and K.H. Robèrt. 2000. Backcasting from non-overlapping sustainability principles: A framework for strategic planning. Int. J. Sust. Dev. World Ecol. 7(4):291–308.

Jabareen, Y. 2008. A new conceptual framework for sustainable development. Environ. Dev. Sustain 10(2):179-192.

Kantabutra, S. 2010. What do we know about vision? Pp. 258-269 in Leader Organizations: Perspectives for a New Era, 2nd Ed., G. Robinson Hickman, ed. Thousand Oaks, CA: Sage.

Marshall, J.D., and M.W. Toffel. 2005. Framing the elusive concept of sustainability: A sustainability hierarchy. Environ. Sci. Technol. 39(3):673-682.

Natural Step. 2011. Backcasting. Natural Step [online]. Available: http://www.naturalstep.org/backcasting [accessed Apr. 19, 2011]

NRC (National Research Council). 2006. Linking Knowledge with Action for Sustainable Development: The Role of Program Management. Washington, DC: National Academies Press.

O'Connell, D., K. Hickerson, and A. Pillutla. 2011. Organizational visioning: An integrative review. Group Organ. Manage. 36(1):103-125.

OECD (Organisation for Economic Co-operation and Development). 1999. European Principles for Public Administration. Sigma Papers 27. Organisation for Economic Co-operation and Development [online]. Available: http://www.oecd.org/dataoecd/26/30/36972467.pdf [accessed Apr. 19, 2011].

OECD. 2001. OECD Environmental Strategy for the First Decade of the 21st Century. Organisation for Economic Co-operation and Development, May 21, 2001 [online]. Available: http://www.oecd.org/dataoecd/33/40/1863539.pdf [accessed Apr. 18, 2011].

OECD. 2011a. OECD Guidance on Developing Safety Performance Indicators: Communities. Organisation for Economic Co-operation and Development [online]. Available: http://www.oecdsafetyindicators.org/node/565 [accessed May 31, 2011].

OECD. 2011b. Sustainable Development Glossary. Organisation for Economic Co-operation and Development [online]. Available: http://www.oecd.org/glossary/0,3414,en_2649_37425_1970394_1_1_1_37425,00.html [accessed Apr. 19, 2011].

OMB (Office of Management and Budget). 2010. Preparation and Submission of Strategic Plans, Annual Performance Plans, and Annual Program Performance Reports. Part 6 in OMB Circular A-11. Preparation, Submission, and Execution of the Budget. Office of Management and Budget, Washington, DC. July 2010 [online]. Available: http://www.whitehouse.gov/sites/default/files/omb/assets/a11_current_year/a_11_2010.pdf [accessed Apr. 19, 2011].

P&G (Proctor & Gamble). 2011. Sustainability. Proctor & Gamble [online]. Available: http://www.pg.com/en_US/sustainability/overview.shtml [accessed June 16, 2011].

Parrish, B.D. 2010. Sustainability-driven entrepreneurship: Principles of organization design. J. Bus. Venturing 25(5):510-523.

Porritt, J. 2007. Capitalism as If the World Matters. London: Earthscan.

Projasek, R.B. 2003. Scoring sustainability results. Environ. Qual. Manage. 13(1):91-98.

Quinn, B. 2009. Walmart's sustainable supply chain. Pollut. Eng. 41(9):24.

UN/EC/IMF/OECD/WB (United Nations, European Commission, International Monetary Fund, Organisation for Economic Co-operation and Development, and World Bank. 2003. Integrated Environmental and Economic Accounting, Handbook of National Accounting. United Nations, European Commission, International Monetary Fund, Organisation for Economic Co-operation and Development, and World Bank [online]. Available: http://unstats.un.org/unsd/envAccounting/seea2003.pdf [accessed June 15, 2011].

UNCED (United Nations Conference on Environment and Development). 1992. Report of the United
    Nations Conference on Environment and Development, Rio de Janeiro, 3-14 June 1992, Annex
    I. Rio Declaration on Environment and Development. A/CONF.151/26 (Vol. I). United Nations
    General Assembly [online]. Available: http://www.un.org/documents/ga/conf151/aconf15126-
    1annex1.htm [accessed Apr. 15, 2011].
Unilever. 2011. Unilever Sustainable Living Plan. Unilever [online]. Available: http://www.
    sustainable-living.unilever.com/ [accessed June 16, 2011].
USGS (U.S. Geological Survey). 2007. Facing Tomorrow's Challenges: U.S. Geological Survey Science
    in the Decade 2007-2017. Circular 1309. U.S. Department of the Interior, U.S. Geological Survey,
    Reston, VA [online]. Available: http://pubs.usgs.gov/circ/2007/1309/pdf/C1309.pdf [accessed
    Apr. 18, 2011].
USGS. 2010. Aligning USGS Senior Leadership Structure with the USGS Science Strategy. Fact Sheet
    2010-3066. U.S. Geological Survey [online]. Available: http://pubs.usgs.gov/fs/2010/3066/
    support/FS2010-3066.pdf [accessed Apr. 19, 2011].
Weiss, E.B. 1989. In Fairness to Future Generations: International Law, Common Patrimony, and
    Intergenerational Equity. Tokyo: Hotei Publishing.
Zaccaro, S.J., and D.J. Banks. 2001. Leadership, vision, and organizational effectiveness. Pp. 181-
    218 in The Nature of Organizational Leadership: Understanding the Performance Imperatives
    Confronting Today's Leaders, S.J. Zaccaro, and R.J. Klimoski, eds. San Francisco, CA: Jossey-
    Bass, Inc.
Zaccaro, S.J., and R.J. Klimoski, eds. 2001. The Nature of Organizational Leadership: Understanding
    the Performance Imperatives Facing Today's Leaders. San Francisco, CA: Jossey-Bass, Inc.

4

Sustainability Assessment and Management: Process, Tools, and Indicators

ELEMENTS OF SUSTAINABILITY ASSESSMENT AND MANAGEMENT

Embedded in the general Sustainability Framework recommended by the Committee on Incorporating Sustainability in the U.S. EPA is an approach to incorporating sustainability to inform decision making. It is called "Sustainability Assessment and Management" and is illustrated as Level 2 in Figure 4-1. This chapter describes the steps involved in this approach, beginning with a screening evaluation to determine whether to conduct the Sustainability Assessment and Management process and to determine the appropriate level of effort or depth of such an assessment. This step is followed by problem definition and scoping, which includes identification of options, preliminary scoping of the analysis, stakeholder involvement, and opportunities for collaboration. The next section describes a set of analytic tools that can be used in the Sustainability Assessment and Management process. The set of potential tools include risk assessment, life-cycle assessment, benefit-cost analysis, ecosystem-services valuation, integrated assessment models, sustainable impact assessment, environmental justice, and present and future scenario tools. This list is not meant to be comprehensive, nor will all of the tools be useful in all cases. The tools, however, are the types of tools that should be in EPA's sustainability toolbox and are likely to be useful in some instances. Following the discussion of tools, the next topic is how to integrate the Sustainability Assessment and Management process into management and policy decisions. Integration into decision making involves summarizing the major results of the assessment in terms of a trade-off and synergy analysis that highlights impacts on important social, environmental, and economic objectives (Box 4-1). This step is followed by presentation of results to the decision makers.

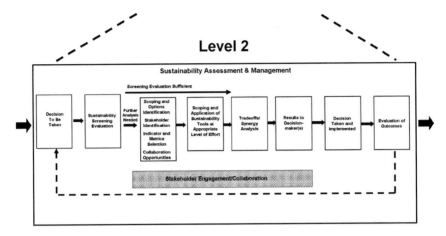

FIGURE 4-1 A framework for EPA sustainability decisions (level 2).

BOX 4-1
Biofuels

Adopting a sustainability framework could help address the social, economic and environmental impacts of biofuel expansion and guide policy decisions toward more sustainable energy supplies. Concerns over energy security, environmental impacts, cost, and availability led to the passage of the Energy Independence and Security Act of 2007 establishing an ambitious goal of producing 36 billion gallons of biofuels annually by 2022. Biofuels are a renewable energy source that can be produced domestically with potentially reduced environmental impacts compared with fossil-fuels. However, the push for biofuels preceded careful sustainability analysis, and the rapid expansion of biofuels production raised its own set of social, economic, and environmental concerns.

The law also requires EPA to report to Congress every three years on the impact of biofuel production on the air, water, and soil quality; ecosystem health and biodiversity, and invasive and noxious plants. The reports are required to include a quantitative assessment of significant environmental changes associated with biofuels production. To date, EPA has not been able to complete a quantitative risk assessment of biofuel production because of a number of factors, including the significant data limitations, substantial uncertainties associated with the production and conversion of biomass feedstocks to biofuels, and a lack of consistency in biofuel production by region.

Impact on food prices: In 2010, 38% of the U.S. corn harvest went to ethanol production. In 2010, total U.S. ethanol production was 13.23 billion gallons (RFA 2011) while U.S. corn production was 12.45 billion bushels (USDA 2011). To produce 13.23 billion gallons, assuming 2.8 gallons of ethanol per bushel of corn, requires 4.725 billion bushels, or 37.95% of total corn production. Some

Finally, once decisions are taken and implemented there should be follow-up evaluation of outcomes on important dimensions of sustainability.

The Sustainability Assessment and Management process should incorporate certain key features:

1. *Comprehensive and systems-based:* Analysis of alternative options should include an integrated evaluation of the social, environmental, and economic consequences.
2. *Intergenerational:* The long-term consequences of alternatives should be evaluated in addition to the more immediate consequences.
3. *Stakeholder involvement and collaboration:* Stakeholders should be involved throughout the process.

The committee recognizes that the formal Sustainability Assessment and Management process can be quite involved and may require EPA to devote

---

analysts blame recent high prices for corn and other crops, at least in part, on biofuel demand (Runge and Senauer 2007; Mitchell 2008). It is difficult, however, to separate out the impact of others factors on food prices, such as the impact of production costs, including high energy prices, weather-related poor harvests, and commodity speculation.

Commercial viability: Biofuel production has been assisted by generous tax credits to refiners, currently $0.45 per gallon for corn ethanol and $1.01 per gallon for cellulosic ethanol, to make it commercially viable (PEW 2009).

Environmental impact: A potential benefit of biofuels is lower life-cycle green-house gas emissions (Farrell et al. 2006; Hill et al. 2006; Wang et al. 2007). Yet, if biofuel expansion causes conversion of forests, wetlands, or native grasslands to croplands, the carbon debt from land-use change can take decades to centuries to repay (Fargione et al. 2008). Increased biofuel production can put pressure on local water supplies and may lead to declines in regional water quality (NRC 2008a). Also, concerns about impacts of changes in land use include the potential negative impacts associated with the expansion of biofuel production on marginal lands and withdrawal of the land from the Conservation Reserve Program (NRC 2010). Biofuel production can also cause an increase in air pollution (Hill et al. 2009).

A 2008 NRC workshop summary on this topic noted that future efforts in this area could include "creating a framework for assessing bioenergy production and biorefineries in the context of sustainability" (NRC 2008b, p.33). Both the United States (EPA 2010) and the European Union (CEU 2010) have requirements to conduct life-cycle assessments of biofuels, but this requirement has to date focused primarily on greenhouse-gas emissions and land-use change rather than the full suite of social, environmental, and economic impacts.

significant staff time and resources to the task. A formal sustainability analysis could also take an extended time period to complete. Therefore, it is important that EPA carefully match the level and depth of the analysis with the scale and magnitude of consequences of the decision at hand. The Sustainability Assessment and Management process should be undertaken for major decisions that could have large impacts on multiple pillars of sustainability. Such an in-depth analysis should not be undertaken for routine or minor decisions, but a systematic approach for addressing sustainability for such decisions could be desirable. The challenge is to match the intensity, detail, and scope of the assessment and management process to the decision needs. This point is discussed further in the screening evaluation section below.

Screening Evaluation

EPA has the discretion to decide what kinds of activities or actions to address in the Sustainability Assessment and Management process. Application of sustainability assessment tools, such as the risk assessment, life-cycle assessment, benefit-cost analysis, ecosystem services valuation, integrated assessment models, sustainability impact assessment, and environmental justice tools described in this chapter, can be applied to programs, policies, and projects; however, not all of them will necessarily require the application of these tools. The agency may wish to focus on major new rules, on complex and important emerging issues, or on making changes to relatively routine decision-making processes. The committee explicitly recommends that EPA develop a sustainability screening approach. There are examples of screening tools used by other governments and the private sector, but EPA will probably need to develop its own set of screening tools.

The screening approach would first determine whether to undertake the Sustainability Assessment and Management approach for any particular program, policy, or project. If it is determined that this process should be undertaken, the screening tool could also provide some guidance on the appropriate analytical tools to apply and on the appropriate degree of depth and detail of analysis needed.

The screening tool should help EPA managers determine whether the full Sustainability Assessment and Management approach is needed. At the one extreme, narrow routine decisions may affect small geographic areas, such as the tens of thousands of permitting decisions on water effluent and air emissions that the agency makes or facilitates annually. For these types of decisions, routine processes have been established. It would be impractical and unworkable to make each of these types of decisions using the formalized Sustainability Assessment and Management approach. Instead, practices and guidelines could be changed so that over time the outcomes are more in line with agency sustainability goals. At the other extreme, the decision-making case may be fairly unique and have wide impact, such as whether to embark on a particular fuel strategy. Such policy decisions will have high impact for all three pillars—social, environmental, and economic—and involve a variety of statutes. Such decisions would probably benefit from the Sustainability Assessment and Management process, either led by EPA or other agencies where EPA has input. EPA may not be the lead federal agency but may be a collaborator, perhaps having an important role in articulating the health and environmental impacts. EPA may have a limited ability to affect the overall decision-making process but may be able to contribute adequately to consideration and analysis of the social/health and environmental dimensions, and potentially voice ways to approach consideration of trade-offs. Finally, other cases may involve repeated but wide-impact decisions (NRC 1996), such as a major expansion of a large refinery, the siting of a power plant, the re-registration of a major use pesticide, actions to address environmental justice issues in a heavily affected community, or a major new rule under the Clean Air Act. These types of high-stakes decisions have potentially large impacts on each of the pillars. They can pose a challenge for the analysis and process. Although any particular new problem may be similar to a previously addressed one, the new problem will likely be sufficiently different to require tailoring of the analysis or process to the specifics of the new problem. In addition, high-stakes decisions typically involve a varied group of interested parties with unequal impacts in terms of those that bear the burden versus those that benefit.

Not all applications of the sustainability assessment tools need to be done at the same level of depth and detail. The distinctions made in administration of the environmental review process under the National Environmental Policy Act (NEPA) provide an example of adjusting the depth of the analysis to the scale of the problem. In addition to providing for categorical exclusions, the NEPA process provides for environmental assessments resulting in a finding of no significant impact (FONSI) or an environmental impact statement (EIS), requiring a much more elaborate analysis and review process (Council on Environmental Quality [CEQ]). (NEPA Regulations, 40 CFR Pt. 1501 [1978]). Varying assessments in the scope and depth of analysis according to the action being considered has long been practiced in the field of risk assessment. A matching of the assessment process to meet the needs of the decision is often recommended as a way to improve the decision-making process (NRC 1996, 2007; IOM 2009). EPA's task is to incorporate sustainability factors and tools—at an appropriately selected level of detail—into existing or new decision-making frameworks so that a multidisciplinary, systematic, and long-term look at the three pillars of sustainability is assured.

Screening is particularly important to avoid undue delays in taking action in the face of environmental problems. A quick scan process can be applied to determine the need for sustainability assessment tools. The quick scan process can determine whether the project is sufficiently large to establish a presumption of possible impacts on one or more pillars of sustainability, to determine the range and magnitude of potential impacts, and to identify which pillars will have large potential impacts. When impacts are likely to be small and the Sustainability Assessment and Management process is not needed, then a library of best-practice techniques and technologies should be consulted and compared with the proposal.

Check lists or impact matrices are often used for screening purposes. The program or project initiative can be broken down into a number of components that can be assessed against social, environmental, and economic criteria of sustainability. For example, in the Swiss assessment process, screening is based on a number of preset social, environmental, and economic criteria (OECD 2010). A rough judgment is made about the causal relationships between the project and the various dimensions of the criteria, and then relevance scores ranging from 0 to 3 are assigned without regard to whether they are positive or negative. A judgment is made on whether there are moderate impacts and potential conflicts between at least two of the pillars (OECD 2010). If both of those conditions are met, then further analysis is needed. How to integrate results from the sustainability screening is discussed further below.

Problem Definition and Planning and Scoping

EPA is engaged in a wide variety of activities as part of its statutory mandates and its initiatives to protect human health and the environment. Specific problems outside EPA's usual activities can also arise, for example, through congressional action, requests for assistance from state or local governments, acts of nature, or terrorism. At the early planning and scoping stage, project managers and analysts diagnose the issue or problem to be addressed. Upfront review of the nature of the problem, credibility of the science, and the decision and legal context helps in considering the nature of the assessment and decision process (Goldstein 1993; NRC 1996, 2007) and whether to embark on a formal or semiformal Sustainability Assessment and Management approach.

An important early step in the process is to identify alternative decisions that could be made (options identification) and to scope the important social (including health), environmental, and economic pillars that could be potentially affected by the decision.

Once attention has been applied to problem definition and identification of alternative options, managers and analysts can begin to develop provisional approaches for the assessment process and the analysis. The Sustainability Assessment and Management approach should begin to develop provisional plans about the level and depth of analysis; the level, extent, and timing of stakeholder engagement; indicators by which they will judge the decision outcomes and process; and collaborative opportunities to explore the range of potential solutions and approaches. To be successful, the overall sustainability process will probably involve a high degree of collaboration, including federal partners, state and local governments, as well as the private sector, nongovernmental organizations, and other stakeholders (NRC 1996, IOM 2009). The levels of information gathering, analysis, and stakeholder involvement for actions that are made subject to the Sustainability Assessment and Management approach will vary depending on the significance of the action and the needs of the decision process (NRC 1996), as discussed in the screening section above.

Another component of the problem definition and scoping process is to select indicators and associated metrics by which to judge success. These metrics can focus on accountability at varying levels of detail and can be directed toward different organizational levels, for example, (1) individual management units within the agency (metrics to show progress toward sustainability goals for program or regional offices), (2) Office of Research and Development (ORD) (a focus on metrics to assess whether the research funded portfolio for ORD is leading to more sustainable solutions to environmental problems), (3) EPA in general, and (4) multiagency collaborations or the United States as a whole (metrics of sustainability regarding overall "performance" of the United States or even the world).

Application of Sustainability Tools

To incorporate sustainability effectively within EPA and to achieve external adoption in various sectors, EPA will have to make use of a variety of assessment tools. EPA will need to develop a set of tools or models that can be used to quantify impacts on important, social, environmental, and economic indicators that might be affected by the program, policy, or project under evaluation. Such tools can provide a uniform and transparent basis on which to evaluate alternatives. The broadening of the analysis from environment and human health to sustainability means that instead of or in addition to risk assessment, additional economic and social factors will need to be considered. This process also means that EPA will need to adopt, develop, or modify a set of tools to conduct such analyses that go beyond traditional risk assessment.

A large number of tools can be applied to address component parts of an analysis. Typically a comprehensive analysis will require the application of a suite of tools. Several principles are important in applying this suite of tools:

- No single tool is likely to be comprehensive; a comprehensive analysis will probably require application of a suite of tools to analyze impacts on social, environmental, and economic pillars of sustainability.
- The suite of tools should include dynamic analysis that analyzes the consequences of alternative options through time (intergenerational component).
- Tools should be capable of delivering quantitative assessments of impacts to the greatest extent feasible.
- It is desirable to have relatively transparent methods that can be easily explained and where the results of the analysis can be effectively communicated to decision makers.

- Data availability will, in part, determine the necessary tool.
- Uncertainty and sensitivity analysis will be required.

Overview of Selected Sustainability Tools

A large number of existing tools can be usefully applied in the Sustainability Assessment and Management process. A small subset of the most appropriate tools, including risk assessment, life-cycle assessment, benefit-cost analysis, ecosystem services valuation, integrated assessment models, sustainability impact assessment, environmental justice tools, and present and future scenario tools are described below. This list is not intended to be a comprehensive list of potential tools but rather a brief review of some important assessment tools.

Risk Assessment

Risk assessment is a tool widely used for characterizing the adverse human health and ecologic effects of exposures. Classically, risk assessments for human health endpoints involves four major steps: a hazard identification, dose-response assessment, exposures assessment and risk characterization (NRC 1983, 1994, 2009; EPA 2005). In the hazard identification step a determination is made about the type of effects potentially caused by the environmental exposure. In the dose response step, the level of exposure such as dose or air concentration is related to the level of adverse effect, such as the incidence of a health effect from an environmental exposure. The exposure assessment characterizes elements of the exposure, for example its intensity, frequency, and timing. The risk characterization combines the dose response and exposure assessments to produce descriptions of the risk for the variety of adverse effects determined in the hazard identification step. In this last step, the uncertainty in the description is also characterized along with variability of the effects in those exposed. For example, a risk assessment may include predictions of the increased incidence of cancer from an environmental chemical exposure in the general population or highly exposed groups or of the margin between the environmental exposure and that causing a noncancer effect seen in the laboratory or in human studies. Ecologic risk assessments evaluate the likelihood that ecologic effects result from environmental exposures to chemicals and other stressors (EPA 1998a). EPA has numerous documents that provide guidance, explain practice, and give operational approaches for specific programs to conduct human health and ecologic risk assessments (EPA 1991, 1996, 1998a, 2000, 2005).

A wide variety of analytic approaches and tools are used in conducting a risk assessment. Risk assessments are important inputs into the process of establishing environmental regulations, cleanup levels, and permitting industrial facilities. An important consideration in any sustainability action is whether environmental or human health will be better or worse off if an action is taken, both near term and in future generations. It is also important to understand not just the direction but also the magnitude. However, it is not always possible to approach these questions quantitatively. Complexity or lack of knowledge may limit the reliability and usefulness of quantitative risk descriptions, but systematic approaches can produce useful qualitative descriptions that can inform decisions. Early on, the NRC (1983) recognized that risk assessments could not always be quantitative, and most recently, the NRC (2009) emphasized the need for tools for fuller characterizations of cumulative risks, including qualitative ones, that adequately account for the full range of chemical and other stressors, particularly for environmental justice contexts. Such risk descriptions could be useful inputs for sustainability decision making. In addition, risk assessment tools for facilitating green chemistry evaluations are needed as green chemistry will continue to be an important component of mitigating human health and environmental risks (NRC 2005a, b). Chapter 5 contains a more detailed discussion of risk assessment.

Life-Cycle Assessment

Life-cycle assessment is a "cradle-to-grave" analysis (or "cradle-to-cradle" ([McDonough and Braungart 2002]) of environmental impacts from production, use, and eventual disposal of a product. Life-cycle assessments are used to analyze the major environmental impacts of various products, to determine how changes in processes could lower the environmental impact, and to compare the environmental impacts of different products (Blackburn 2007). Life-cycle assessments are already used by EPA and have been used to compare the environmental impacts of transportation fuels and specifically to judge whether biofuels meet requirements for carbon-emissions reductions relative to fossil fuels (EPA 2009). Life-cycle assessments take a systems perspective to include the whole production process, from production of raw materials to eventual disposal and is therefore consistent with, and often an essential component of, sustainability analysis. Life-cycle assessments require a large amount of data on necessary inputs, outputs, and various types of environmental emissions of processes. The availability of standardized economy-wide input-output coefficients for ready use simplifies this challenge. Other challenges with applying life-cycle analysis in a sustainability context involve decisions on where to set system boundaries and what to assume about future technologies.

Benefit-Cost Analysis

Benefit-cost analysis is a widely used tool from economics to evaluate the net benefits of alternative decisions. Benefit-cost analysis seeks to assess the change in welfare for each individual affected by a policy choice, measured in a common monetary metric, under a set of alternatives. Most benefit-cost analyses then aggregate the measure of individual net benefits to find a social net benefit and then rank the alternatives. There have been concerns that benefit-cost analysis as commonly applied to environmental issues places too much emphasis on the economic costs and too little on benefits and their distribution (OECD 2006). Recent developments in benefit-cost analysis as applied to environmental issues can be used in an attempt to ensure that the full range of benefits and costs can be taken into account better. These developments include, for example, integrating life-cycle analysis into benefit-cost analysis, having improved methods of estimating the value of ecosystem services, and paying close attention to distribution of benefit and costs across different groups in society to address environmental justice concerns (Pearce et al. 2006).

Of particular concern for sustainability analysis is the weighting (discounting) of benefits and costs that accrue to future generations compared with those that accrue to the current generation (intergenerational equity). Although discounting will account for the costs to present generations of providing protections, opponents of benefit-cost analysis perceive discounting as inconsistent with an environmental law's forward-looking premise because the standard technique of constant exponential discounting can have a potentially large adverse effect on the perceived benefits—such as protecting against long-latency diseases like cancer—that aim to prevent future harm (Harrington et al. 2009). For further discussions on alternative discounting methods, see Pearce (2006); with specific reference to the use of discounting in climate policy, see Nordhaus (2007) and Stern and Taylor (2007). Such issues will need to be addressed in sustainability analyses that use benefit-cost analyses.

Ecosystem Services Valuation

Ecosystem services are goods and services that contribute to human well-being and are generated by ecosystem processes. For example, ecosystems can filter contaminants to provide clean water for human use and modulate water flow, reducing the probabilities of flooding and providing higher flows during drier periods. Ecosystem-service valuation is an attempt to measure the relative benefits of ecosystem services in a common metric (usually a monetary metric). Ecosystem-services valuation requires integration of ecological and other natural sciences (EPA SAB 2009). It is used to better understand the provision of services as a consequence of the state of the ecosystem ("ecologic production functions") along with economics and other social sciences to gain an understanding of how nature contributes to human well-being ("valuation").

Ecosystem-service valuation measured in money terms can be used in benefit-cost analysis to capture a more complete picture of the net benefits of alternative actions. Economic valuation methods for ecosystem-service valuation are well described in both NRC (2005a,b) and EPA SAB (2009). EPA SAB (2009) also reviewed a number of other noneconomic approaches to valuation. For sustainability analysis, what is of most interest is how the value of ecosystem services will probably change through time. Notions of sustainability can be assessed through an evaluation of the value of natural capital and other forms of capital assets (including manufactured capital, human capital, and social capital). The value of natural capital is the contribution of an attribute of an ecosystem to present value of the flow of services through time. de Groot et al. (2002) also provided a conceptual framework and typology for the classification, description, and valuation of ecosystem goods, functions, and services.

Integrated Assessment Models

Integrated assessments cross disciplinary lines to merge theory and data from multiple disciplines to address complex environmental issues. Modeling is the standard tool used for conducting an integrated assessment. Integrated assessment models, such as the Global Change Assessment Model (GCAM), arose in the study of climate change, bringing together global circulation models and economic models to assess the probable benefits and costs of alternative energy- and climate-policy choices (Hannam et al. 2009). Although typically not called integrated assessment models, models used for ecosystem-services valuation are also examples that integrate models from multiple disciplines to assess the benefits and costs of alternative policy choices. The strength of integrated assessments is that they combine knowledge from multiple disciplines needed to understand how human actions might affect the system in important ways (e.g., greenhouse gas emissions and the climate system). Integrated assessments often take an expansive and long-term view, which is suitable for sustainability analysis. Integrated assessment models are often complex, tending to make them nontransparent to nonexperts. Furthermore, outcomes can be sensitive to modeling assumptions for that might have inadequate factual bases for clearly determining the right assumption to use. Still, integrated assessment models will often be needed to understand the relationships among the social, environmental, and economic pillars of sustainability in the context of a particular decision.

Sustainability Impact Assessment

Sustainability impact assessment is used to analyze the probable effects of a particular project or proposal on the social, environmental, and economic pillars of sustainability. This assessment is also used to develop integrated policies that "take full account of the three sustainable development dimensions" and include the "cross-cutting, intangible and long-term considerations" of those policies (OECD 2010). Sustainability impact assessment is used in many European countries and in Canada but has not been used to any great extent in the United States (Zerbe and Dedeurwaerdere 2003). Sustainability impact assessment is modeled on, but different from, environmental impact assessment, which was pioneered in the United States through the National Environmental Policy Act of 1969 and is now widely used around the world. Environmental impact assessment tends to focus primarily on the projected environmental effects of a particular action and alternatives to that action. The purpose of environmental assessment is to ensure full consideration of environmental impacts and alternatives, with the understanding that such consideration will ordinarily (but not necessarily) reduce the environmental impact of the decision. The objective of the sustainability impact assessment, in contrast, is not only to minimize the environmental impact but also to optimize a particular decision's contribution to sustainability (Gibson 2005).

Environmental Justice Tools

Environmental justice tools are analytic methods for judging whether communities are experiencing inordinately high environmental and health burdens and for evaluating the sustainability of communities. The tools include quantitative and semiquantitative methods for screening communities of concern, for conducting specific community evaluations of cumulative environmental impacts or risks, and for looking at cumulative exposures and impacts in planning for land use (OEHHA 2010). An approach for assessing inequalities in environmental exposures has also been developed similar to one used to assess income distribution inequities (Su et al. 2009).

Environmental justice tools include guidance documents for working with communities to engage in problem-solving efforts (EPA 2008) and evaluating whether environmental justice concerns have been adequately addressed in an assessment (e.g., EPA 1998b). The goal of a cumulative risk assessment in a community setting is to fully account for the combined effects of multiple exposures—chemical, biologic, psychosocial, and physical—on a community, a goal that cannot be achieved using standard risk assessment methodology (IOM 2009). This goal has resulted in a call for simple tools to adequately address community concerns in evaluating community status with respect to environmental justice. Environmental justice and cumulative impact analyses can be used in priority-setting processes to direct resources to address the most heavily affected communities, to evaluate equity and fairness issues in siting and permitting decisions, and to facilitate community considerations of resource use (Morello-Frosh et al. 2011). In sustainability decision making, environmental justice tools may be similarly used.

Present Conditions and Future Scenario Tools

The Sustainability Assessment and Management approach requires an evaluation of present and future conditions to show that present decisions and actions are not compromising future human and ecologic health and well-being. A requirement of these elements, therefore, is to be able to forecast potential future conditions as a function of the decision option chosen. The forecast should take into account both the decision options and the underlying biophysical, social, and economic forces that will influence system dynamics.

Forecasting conditions relevant to sustainability poses a paradox. There is no standardized universally accepted way to forecast the future and, in fact, most forecasts are wrong to a greater or lesser degree. On the other hand, almost all policy actions are premised on assumptions about future conditions. Forecasting is unavoidable when dealing with sustainability, but our ability to do forecasting is limited. At a minimum, this means that the forecasted premises needs to be made explicit and the uncertainty of the forecast also needs to be explicit because the robustness of the sustainability assessment will depend on the degree of uncertainty of the forecast.

One standard approach to specifying future conditions is to use scenario analysis (Schwartz 1991). A scenario is a plausible story about how the future might unfold from current conditions given assumptions about biophysical processes, human behavior, policy, and institutions. Major global assessments, the "Global Environmental Outlook" (UNEP 2002), the "Special Report on Emissions Scenarios" (SRES) of the IPCC (Nakicenovic et al. 2000), the "Millennium Ecosystem Assessment" (MEA 2005), and the OECD Environmental Outlook to 2030 (OECD 2008) have each generated scenarios of future conditions. Scenarios are useful in situations of great complexity and uncertainty, as is the case in global assessment of complex systems. Creative thinking about a set of scenarios can highlight the potential range of plausible future outcomes. Scenarios, such as a range of scenarios about land-use change, can be used at local or regional levels. See Box 4-2 for an example of such a scenario for global biodiversity.

BOX 4-2
Scenarios for Global Biodiversity

Quantitative scenarios used to evaluate the impact of future socioeconomic development pathways and ecosystems services have indicated that biodiversity will continue to decline over the twenty-first century (Pereira et al. 2010). However, Pereira et al. (2010) noted that the "range of projected changes is much broader than most studies suggest partly because there are significant opportunities to intervene through better policies" (p. 1496). In this model, scenarios of socioeconomic development pathways modeled include population growth, fossil fuel use, and food demand. The projections of direct drivers for this include climate change, land-use change, water extraction, and fish harvesting pressure. The projections of impacts on biodiversity are twofold: habitat or functional group-level changes and species-level changes. Finally, the projections of impacts on ecosystem services include impacts on provisions, regulation, support, and cultural services. The authors noted that to better inform policy, "scenarios must move beyond illustrating the potential impacts of global change on biodiversity toward more integrated approaches that account for the feedbacks that link environmental drivers, biodiversity, ecosystems services, and socioeconomic dynamics" (Pereira et al. 2010, p. 1501).

EPA needs to improve its forecasting ability both in the context of the Sustainability Assessment and Management approach and, more generally, needs to be better able to anticipate and deal with future environmental problems. As stated in a previous NRC report (2005a, p.8), "Federal environmental agencies should undertake an assortment of research initiatives to collect, appraise, develop, and extend analytical activities related to forecasting in order to improve environmental understanding and decision making."

Trade-off and Synergy Analysis

Trade-off and synergy analysis is a fundamental component of the Sustainability Assessment and Management approach. The objective is to maximize synergies (social, environmental, and economic benefits of a decision) and to minimize the adverse effects of conflicts among the three pillars. Because conflicts lead to trade-offs among the three pillars and because improperly managed trade-offs can compromise environmental protection, public health, or other key aspects of sustainability, clear trade-off rules are required. Most basically, "trade-off decisions must not compromise the fundamental objective of net sustainability gain" (Gibson 2006, p.175). OECD recommends,

> "Preference should be given to those scenarios in which none of the three sustainability dimensions is too strongly impaired. The proposed options should all meet the following minimum requirements: (i) environmental standards established to protect human and environmental health; and (ii) living standards in keeping with social well-being or to safeguard human rights. The aim is to develop "win-win" situations where mutually-reinforcing gains can strengthen the economic base, ensure equitable living conditions, and protect and enhance the environment. Where this is impossible, the trade-offs should be clearly indicated to guide decision makers" (OECD 2010).

It will be important for EPA to establish a systematic way to analyze and quantify alternatives. One approach to analyzing conservation and management alternatives was used in the application of spatially explicit models of multiple ecosystem services and biodiversity conservation (Polasky et al. 2011). These models illustrated how predictions could be used to analyze alternative conservation and management strategies, and by comparing maps of ecosystem services and biodiversity, decision makers could identify areas that provide high levels of both. Outcomes compared across different management alternatives give insight into which alternatives are best. The analysis can be used to identify new strategies that may improve results for key ecosystem services or biodiversity conservation objectives. Gibson (2006) provides guidelines for approaching trade-off analysis:

> *Maximum net gains:* Any acceptable trade-off or set of trade-offs must deliver net progress towards meeting the requirements for sustainability; it must seek

mutually reinforcing, cumulative and lasting contributions and must favor achievement of the most positive feasible overall result while avoiding significant adverse effects.

Burden of argument on trade-off proponent: Trade-off compromises that involve acceptance of adverse effects in sustainability-related areas are undesirable unless proven (or reasonably established) otherwise; the burden of justification falls on the proponent of the trade-off.

Avoidance of significant adverse effects: No trade-off that involves a significant adverse effect on any sustainability requirement area (for example, any effect that might undermine the integrity of a viable socio-ecological system) can be justified unless the alternative is acceptance of an even more significant adverse effect.

Protection of the future: No displacement of a significant adverse effect from the present to the future can be justified unless the alternative is displacement of an even more significant negative effect from the present to the future.

Explicit justification: All trade-offs must be accompanied by an explicit justification based on openly identified, context-specific priorities as well as the sustainability decision criteria and the general trade-off rules.

Open process: Proposed compromises and trade-offs must be addressed and justified through processes that include open and effective involvement of all stakeholders.

The above examples of evaluating trade-offs are illustrative of the type of analysis EPA will need to undertake. The committee expects that EPA will adopt trade-off rules that are consistent with its existing legal authority and that are based on consideration of trade-off rules used elsewhere. The committee also expects that these trade-off rules may change over time as EPA gains greater experience with their use.

Communication of Results to Decision Makers

Following scoping and options identification, screening evaluation, application of tools, and trade-off and synergy analysis, communication of results will need to be integrated into the decision-making process at a point when the information can be considered in formulating the policy or program or taking a major action subject to sustainability review. Results should be available as early as is practicable; an assessment may show the need for further information or action on particular issues. The roles of the decision makers and the offices contributing expertise will need to be defined and accountability measures should be in place to ensure that the decision maker gives due consideration to the results of the assessment in acting on the subject.

Decisions Taken and Implemented

A series of briefing documents at a depth appropriate to the decision would probably be prepared to inform the decision making. The range of options and the associated social (including health), environmental, and economic impacts for each option would be presented along with any trade-off analyses that may have been undertaken. As is the case with alternatives analysis under NEPA, options in a sustainability assessment allow the decision maker to understand different ways of taking a particular action and thus provide the decision maker with more choices to reduce adverse impacts. In addition, options in a sustainability assessment allow the decision maker to find better ways of improving social well-being, economic development, and environmental protection at the same time. Options also make clearer the causes of any trade-offs among the three pillars and help the decision maker to reduce the adverse effects of any trade-offs. Questions could arise that would require some additional analysis as well as follow-up with various stakeholders and collaborators.

Evaluation of Outcomes

An important component in communicating the potential benefits of adopting sustainability and justifying further efforts directed toward sustainability is demonstrating the effectiveness of prior actions and providing the information to be used in the feedback loop to modify goals. Such work goes by such names as project evaluation, post facto evaluation, or accountability analysis (NRC 2005b). Evaluation is useful in identifying best practices, reducing uncertainties, and identifying additional linkages. Evaluation, like tools and indicators, is founded on having the appropriate data. There is a significant literature on program evaluation, including methods for measuring program performance, including but not limited to Cronbach 1980, Chelimsky 1997, Vedung 1997, Stufflebeam 2001, and Posner 2004.

At the first level, evaluation should compare the observed response of the indicator (or associated metric) to the project goals. Discrepancies should be evaluated to identify weaknesses in the assessment process, including the tools and data used. This evaluation can be assisted by identifying additional data to better characterize system linkages and responses in indicators other than those that are used to assess goals. An important question to address is whether the response was within the range of uncertainty estimated during the alternative options analysis.

Planning for evaluation includes identifying additional data and tools that are critical in understanding the system at a level that the predictability of future similar projects or policies is improved. Given the transgenerational nature of many sustainability indicators and goals, the evaluation may need to be based on indicators that are longer term than those directly used in assessing how well a project or policy meets the stated goals.

Identifying the most appropriate indicators can be driven by models; for example, sensitivity analysis can be used to quantify how system indicators will respond to perturbations (e.g., policies and projects). The dual role of sustainability indicators is to both measure and communicate the range of factors that are involved in the decision-making process. Indicators, generally, are measures of the system of interest and can be either directly observed or derived quantities.

Sustainability Indicators

Substantial work has been done on the subject of sustainability indicators internationally (see Hak et al. [2007] for a review of the state of the art in sustainability indicators), much of which EPA has been involved in. The corporate sector has also developed indicators and metrics for sustainable performance (Székely and Knirsch 2005). EPA's new 5 year strategic plan calls for the development of additional indicators that will improve understanding of the integrated and complex relationships involved in maintaining human health and environmental well-being (EPA 2010). The plan envisions that the development of additional indicators will be useful in tracking changes in environmental justice, children's health, and regional ecosystems, such as the Great Lakes. The plan also foresees development and use of indicators to advance the sustainable communities program. In preparation for EPA's next report on the environment, a task force has begun work to identify indicators of sustainability and associated metrics.

In general, the work on sustainability indicators has tended to rely on work already done to gather social, environmental, and economic data. Most sustainability indicators are transformations of these data, and the experts involved in the sustainability-indicator efforts have been guided by a need, not only for relevance, but also for practicality. The committee recognizes the need for EPA to identify indicators and indicator sets that can help it to take the opportunities that a sustainability approach presents both locally and globally (Box 4-3). In serving both a measurement and a communication role, indicators can be used to promote beneficial change and also identify potential threats to sustainability. Sustainability indicators differ from those developed to measure a specific outcome of a program, such as an air quality parameter, as they must be able to capture information across multiple factors. Sustainability indicators would synthesize and report on various complex areas, including social, environmental, and economic aspects. For example, a well-known indicator for assessing health and well-being in developing countries is infant mortality, as this indicator can be used singularly to infer information about maternal health, behaviors, and economic conditions in a particular country. A sustainability indicator would also be actionable in that the agency can take practical steps to address factors contributing to an indicator to attain sustainability goals. Although some sustainability challenges addressed in one region may overlap to some degree in another, there will also be distinct challenges in that region and, to that end

---

BOX 4-3 Indicator Attributes

- **Actionable**—practical steps taken to address factors contributing to an indicator to attain sustainability goals
- **Transferable and scalable**—adaptable at regional, state, or local levels
- **Intergenerational**—The fair distribution of costs and benefits among different generations
- **Definable**—simple to understand and easily communicated
- **Relevant**—be relevant to actual or anticipated policies
- **Important**—reflect an important aspect of the social, environmental, or economic pillars
- **Measureable**—measure something of obvious value to the public and decision makers
- **Durable**—long-term relevance

---

sustainability indicators would be transferable and scalable and be adaptable at a regional, state, or local level.

Because some sustainability goals may require long-term solutions, sustainability indicators would be applicable in the short-term but also intergenerational and usable in a long-term time frame. Inherent in sustainability is concern about intergenerational impacts, thus differentiating sustainability indicators from many commonly used environmental indicators that reflect the current state of the environment. One approach that can be used to address intergenerational dimensions is the use of "stock-and-flow indicators." Stock-and-flow indicators address the availability of a resource and the rate of depletion or growth, and are thus more intergenerational; policy indicators are more applicable to assessing change over short periods of time (intragenerational) that can be attributed to policies. Use of stock-and-flow indicators will require multiagency cooperation. An issue with the stock-and-flow indicators is their complexity, and as such, their development is more difficult (NRC 1999). Assessing impacts across generations can complicate the quantification of an indicator and introduce additional uncertainty. Thus, one component of quantifying an indicator will also be assessing the related uncertainty. Further discussion of indicators can be found in Appendix E.

Reporting

EPA would benefit from systematically conducting and publishing results of sustainability evaluations of major decisions, projects, activities, and programs by using indicators that provide accurate, comprehensive, and reliable information. Stakeholders could be further engaged by publicizing the results of these evaluations, including not only successes but also lessons learned and areas where data are insufficient to draw a conclusion. Decision makers need to assure that proper stakeholder engagement has occurred and is part of final decisions. Following implementation, the effectiveness of the decision/action as well as verification of the impacts will need to be pursued.

Some states (e.g., Michigan) require analysis and periodic reporting on emerging environmental and sustainability issues.[1] Such reporting can serve as a kind of early warning system and enable public and private decision makers to address issues at an earlier stage than might be possible otherwise. EPA would benefit from this kind of analysis and reporting as part of future reports on the environment. EPA could also benefit from the practice of systematically documenting and providing public information about the sustainability co-benefits of its actions and decisions, including not only environmental benefits but also economic and social benefits. The object of this practice would be to educate the public about the links between environmental protection and human well-being, and to help the public understand the role that EPA has played and will continue to play in fostering sustainability. When quantitative description of such benefits is not possible or feasible, qualitative description of these benefits would be appropriate.

The agency may wish to consider, at the regional and headquarters levels, regularly producing a sustainability report utilizing widely recognized indicators (such as the "Global Reporting Initiative"[2]). Consistent with Executive Order 13514,[3] EPA would benefit from implementing an internal agency program to identify key sustainability indicators, implementing a tracking and reporting system to demonstrate progress toward the goals of more sustainable operational practices and benchmark performances against other federal or government agencies and private sector organizations. The agency is already required to report on the seven metrics of sustainability and energy performance described in the Executive Order and recently produced a FY2010 OMB Scorecard on Sustainability/ Energy to document its performance (EPA 2011).

Stakeholder Engagement and Collaboration

Stakeholder engagement is generally cited as one of the essential elements of a sustainability approach (Feldman 2002). The Sustainability Framework outlined in this report contemplates that EPA will involve stakeholders at appropriate times throughout the process. The Sustainability Assessment and Management

---

[1] The Michigan Environmental Indicators Act (P.A. 1999, No. 195); Codified at Mich. Comp. Laws Ann. 324.2521.

[2] The GRI is a "network-based organization that pioneered the world's most widely used sustainability reporting framework. The Reporting Framework sets out the principles and Performance Indicators that organizations can use to measure and report their economic, environmental, and social performance" (GRI 2011).

[3] Executive Order 13514, titled Federal Leadership in Environmental, Energy, and Economic Performance, sets sustainability goals for Federal agencies, including the need for improvements in environmental, energy and economic performance.

approach has as a component the identification of stakeholders interested in a particular program or action during the scoping process after a decision has been made to perform some level of sustainability assessment. EPA has extensive experience with public participation activities, many of which are legal mandates, in its regulatory work. The sustainability assessments suggested here are not regulatory requirements, and their implementation presents a new opportunity to advance the state of practice of involving people in governmental decision making.

FINDINGS AND RECOMMENDATIONS

4.1. Key Finding: The Sustainability Assessment and Management approach requires application of a suite of tools capable of analyzing the full set of current and future social, environmental, and economic consequences of alternative options. Many tools already exist, and much activity is under way in the United States and globally to develop such tools. Some tools will need modification or expansion to be appropriate and some new tools will need to be developed (p.60-65).

4.1. Key Recommendation: EPA should develop a "sustainability toolbox" that includes a suite of tools for use in the Sustainability Assessment and Management approach. Collectively, the suite of tools should have the ability to analyze present and future consequences of alternative decision options on the full range of social, environmental, and economic indicators. Application of these tools, ranging from simple to complex, should have the capability for showing distributional impacts of alternative options with particular reference to vulnerable or disadvantaged groups and ecosystems.

4.2. Finding: An important step in the Sustainability Assessment and Management approach is an evaluation of present and future conditions to show that present decisions and actions are not compromising future human and ecologic health and well-being. Therefore, a requirement is to be able to forecast potential future conditions as a function of the decision option chosen, although there will always be some degree of uncertainty attached to the forecast (p.64-65).

4.2. Recommendation: EPA should identify potential future environmental problems, consider a range of options to address problems, and develop alternative projections of environmental conditions and problems.

4.3. Finding: The culture change being proposed here will require EPA to conduct an expanding number of assessments. Although EPA has been involved in state-of-the-environment and environmental assessments, it currently does not have a formalized approach to conducting or participating in the analyses required in the Sustainability Assessment and Management approach. Thus, such assessments could readily miss sustainability concerns not typically considered in past environmental assessments, including social and economic issues and environmental justice (p.58-59).

4.3. Recommendation: The agency should develop a tiered formalized process, with guidelines, for undertaking the Sustainability Assessment and Management approach to maximize benefits across the three pillars and to ensure further intergenerational social, environmental, and economic benefits that address environmental justice.

4.4. Finding: Screening is often used in other OECD countries prior to undertaking full sustainability assessments; criteria examined include the magnitude of the activity and potential short-term and long-term conflicts between at least two dimensions of sustainability (p.56).

4.4. Recommendation: EPA should formalize a screening procedure for implementing the Sustainability Framework recommended by the committee.

4.5. Finding: Economic benefit-cost analysis as commonly applied to environmental issues often does not adequately account for the full range of ecosystem benefits, take intergenerational considerations into account sufficiently, or take into account the distribution of benefits and costs among population groups (p.61).

4.5. Recommendation: EPA should continue to adapt its current method of cost benefit analysis for sustainability by, among other things, improving its estimates of the value of ecosystem services, extending its boundaries by incorporating life-cycle analysis, and better addressing intergenerational and environmental justice considerations.

4.6. Finding: Risk analysis as commonly applied to environmental issues often does not adequately account for the full range of human health and ecosystem risks, including cumulative risks, intergenerational considerations, and the distribution of risks among population groups. In addition, better methods are needed to support consideration of health and environmental effects for the green chemistry goal of safer products and more sustainable chemical usage (p.60).

4.6. Recommendation: EPA should develop a range of risk assessment methods to better address cumulative risk and intergenerational and environmental justice considerations and to support comparisons of chemicals as part of an alternatives analysis for green chemistry applications.

4.7. Finding: EPA and other organizations have developed and continue to develop environmental indicators; however, appropriately addressing sustainability in the decision-making process will require additional attention to economic and social issues, including environmental justice (p.69).

4.7. Recommendation: EPA should expand its environmental indicators to address economic and social issues in collaboration with other federal agencies to address economic and social issues, and consider adopting them and developing appropriate metrics to inform sustainability considerations for state and local actors. Where relevant, these indicators should allow for international comparisons and the rapid adoption and adaptation of best practices from other countries responding to the challenges of sustainability.

REFERENCES

Blackburn, W.R. 2007. The Sustainability Handbook: The Complete Management Guide to Achieving Social, Economic, and Environmental Responsibility. Sterling, VA: Earthscan.

CEU (Council of the European Union). 2010. Communication from the Commission on the practical implementation of the EU biofuels and bioliquids sustainability scheme and on counting rules for biofuels. O. J. EU. C160:8-16 [online]. Available: http://eur-lex.europa.eu/LexUriServ/LexUriServ.do?uri=OJ:C:2010:160:0008:0016:EN:PDF [accessed Apr. 19, 2011].

Chelimsky, E. 1997. The coming transformations in evaluation. Pp. 1-26 in Evaluation for the 21st Century: A Handbook, E. Chelimsky, and W.R. Shadish, eds. Thousand Oaks, CA: Sage.

Cronbach, L.J. 1980. Toward Reform of Program Evaluation: Aims, Methods and Institutional Arrangements. San Francisco, CA: Jossey-Bass.

de Groot, R.S., M.A. Wilson, and R.M.J. Boumans. 2002. A typology for the classification, description and valuation of ecosystem functions, goods and services. Ecol. Econ. 41(3): 393-408.

EPA (U.S. Environmental Protection Agency). 1991. Guidelines for Developmental Toxicity Risk Assessment. EPA/600/FR-91/001. Risk Assessment Forum, U.S. Environmental Protection Agency, Washington, DC. December 1991 [online]. Available: http://www.epa.gov/raf/publications/pdfs/DEVTOX.PDF [accessed May 2, 2011].

EPA. 1996. Guidelines for Reproductive Toxicity Risk Assessment. EPA/630/R-96/009. Risk Assessment Forum, U.S. Environmental Protection Agency, Washington, DC. October 1996 [online]. Available: http://www.epa.gov/raf/publications/pdfs/REPRO51.PDF [accessed May 2, 2011].

EPA. 1998a. Guidelines for Ecological Risk Assessment. EPA/630/R-95/002F. Risk Assessment Forum, U.S. Environmental Protection Agency, Washington, DC. April 1998. [online]. Available: http://www.epa.gov/raf/publications/pdfs/ECOTXTBX.PDF [accessed July 13, 2011].

EPA. 1998b. Final Guidance for Incorporating Environmental Justice Concerns in EPA's NEPA Compliance Analyses. U.S. Environmental Protection Agency. April 1998 [online]. Available: http://www.epa.gov/compliance/ej/resources/policy/ej_guidance_nepa_epa0498.pdf [accessed May 2, 2011].

EPA. 2000. Supplementary Guidance for Conducting Health Risk Assessment of Chemical Mixtures. EPA/630/R-00/002. Risk Assessment Forum, U.S. Environmental Protection Agency, Washington, DC [online]. Available: http://cfpub.epa.gov/ncea/cfm/recordisplay.cfm?deid=20533#Download [accessed May 2, 2011].

EPA. 2005. Guidelines for Carcinogen Risk Assessment. EPA/630/P-03/001F. Risk Assessment Forum., U.S. Environmental Protection Agency, Washington, DC [online]. Available: http://www.epa.gov/cancerguidelines/ [accessed May 2, 2011].

EPA. 2008. EPA's Environmental Justice Collaborative Problem-Solving Model. EPA-300-R-06-002. U.S. Environmental Protection Agency, Washington, DC. June 2008 [online]. Available: http://www.epa.gov/compliance/ej/resources/publications/grants/cps-manual-12-27-06.pdf [accessed May 2, 2011].

EPA. 2009. EPA Lifecycle Analysis of Greenhouse Gas Emissions from Renewable Fuels. EPA-420-F-09-024. Office of Transportation and Air Quality, U.S. Environmental Protection Agency. May 2009 [online]. Available: http://www.epa.gov/oms/renewablefuels/420f09024.pdf [accessed May 2, 2011].

EPA. 2010. Fiscal Year 2011-2015 EPA Strategic Vision: Achieving Our Vision. U.S. Environmental Protection Agency, Washington, DC [online]. Available: http://www.epa.gov/planandbudget/strategicplan.html [accessed Apr. 18, 2011].

EPA. 2011. FY2010 OMB Scorecard on Sustainability/Energy. U.S. Environmental Protection Agency [online.] Available: http://www.epa.gov/aboutepa/FY2010OMBScorecard.pdf [accessed Apr. 25, 2011].

EPA SAB (U.S. Environmental Protection Agency Science Advisory Board). 2009. Valuing the Protection of Ecological Systems and Services. EPA-SAB-09-012. Science Advisory Board, U.S. Environmental Protection Agency, Washington, DC. May 2009 [online]. Available: http://yosemite.epa.gov/sab/sabproduct.nsf/WebBOARD/SAB-09-012/$File/SAB%20Advisory%20Report%20full%20web.pdf [accessed June 8, 2011].

Fargione, J., J. Hill, D. Tilman, S. Polasky, and P. Hawthorne. 2008. Land clearing and the biofuel carbon debt. Science 319(5867):1235-1238.

Farrell, A.E., R.J. Plevin, B.T. Turner, A.D. Jones, M. O'Hare, and D.M. Kammen. 2006. Ethanol can contribute to energy and environmental goals. Science 311(5760):506-508.

Feldman, I. 2002. The stakeholder convergence: Public participation and sustainable business practices. Pp. 53- 58 in The New "Public": The Globalization of Public Participation, C. Bruch, ed. Washington, DC: Environmental Law Institute [online]. Available: http://www.eli.org/pdf/PPP/part1chap3.pdf [accessed July 13, 2011].

Gibson, R. 2005. Sustainability Assessment: Criteria and Process. Sterling, VA: Earthscan.

Gibson, R. 2006. Sustainability assessment: Basic components of a practical approach. IAPA 24(3):170-182.

Goldstein, B.D. 1993. Science and an EPA mission statement. Environ. Health Perspect. 101(6):466-467.

GRI (Global Reporting Initiative). 2011. What is GRI? [online]. Available: http://www.globalreporting.org/AboutGRI/WhatIsGRI/ [accessed June 1, 2011.]

Hak, T., B. Moldan, and A.L. Dahl, eds. 2007. Sustainability Indicators: A Scientific Assessment. Washington, DC: Island Press.

Hannam, P., G. Kyle, and S.J. Smith. 2009. Global Deployment of Geothermal Energy Using a New Characterization in GCAM 1.0. PNNL-19231. Pacific Northwest National Laboratory, Richland, WA. September 2009 [online]. Available: http://www.pnl.gov/main/publications/external/technical_reports/PNNL-19231.pdf [accessed June 8, 2011].

Harrington, W., L. Heinzerling, and R. Morgenstern, eds. 2009. Reforming Regulatory Impact Analysis. Washington, DC: Resources for the Future [online]. Available: http://www.rff.org/RFF/Documents/RFF.RIA.V4.low_res.pdf [accessed Apr. 19, 2011].

Hill, J., E. Nelson, D. Tilman, S. Polasky, and D. Tiffany. 2006. Environmental, economic, and energetic costs and benefits of biodiesel and thanol biofuels. Proc. Natl. Acad. Sci. 103(30):11206-11210.

Hill, J., S. Polasky, E. Nelson, D. Tilman, H. Huo, L. Ludwig, J. Neumann, H. Zheng, and D. Bonta. 2009. Climate change and health costs of air emissions from biofuels and gasoline. Proc. Natl. Acad. Sci. 106(6):2077-2082.

IOM (Institute of Medicine). 2009. Environmental Health Sciences Decision Making: Risk Manage-
    ment, Evidence, and Ethics: Workshop Summary. Washington, DC: National Academies Press.
McDonough, W., and M. Braungart. 2002. Cradle to Cradle: Remaking the Way We Make Things.
    New York: North Point Press.
MEA (Millennium Ecosystem Assessment). 2005. Ecosystems and Human Well-Being: Synthesis.
    Washington, DC: Island Press.
Mitchell, D. 2008. A Note on Rising Food Prices: Policy Research Working Paper 4682. World Bank,
    Washington, DC [online]. Available: http://econ.tu.ac.th/class/archan/RANGSUN/EC%20460/
    EC%20460%20Readings/Global%20Issues/Food%20Crisis/Food%20Price/A%20Note%20
    on%20Rising%20Food%20Price.pdf [accessed Apr. 19, 2011].
Morello-Frosch, R., M. Zuk, M. Jerrett, B. Shamasunder, and A.D. Kyle. 2011. Understanding the
    cumulative impacts of inequities in environmental health: Implications for policy. Health Aff.
    30(5):879-887.
Nakicenovic, N., J. Alcamo, G. Davis, B. de Vries, J. Fenhann, S. Gaffin, K. Gregory, A. Grübler,
    T.Y. Jung, T. Kram, E.L. La Rovere, L. Michaelis, S. Mori, T. Morita, W. Pepper, H. Pitcher, L.
    Price, K. Riahi, A. Roehrl, H.H. Rogner, A. Sankovski, M. Schlesinger, P. Shukla, S. Smith, R.
    Swart, S. van Rooijen, N. Victor, and Z. Dadi. 2000. Special Report on Emissions Scenarios.
    A Special Report of Working Group III of the Intergovernmental Panel on Climate Change.
    Cambridge: Cambridge University Press [online]. Available: http://www.grida.no/publications/
    other/ipcc_sr/?src=/climate/ipcc/emission/ [accessed Apr. 19, 2011].
Nordhaus, W. 2007. Critical assumptions in the Stern Review on climate change. Science
    317(5835):201-202.
NRC (National Research Council). 1983. Risk Assessment in the Federal Government: Managing the
    Process. Washington, DC: National Academy Press.
NRC. 1994. Science and Judgment in Risk Assessment. Washington, DC: National Academy Press.
NRC. 1996. Understanding Risk: Informing Decisions in a Democratic Society. Washington, DC:
    National Academy Press.
NRC. 1999. Nature's Numbers: Expanding the National Economic Accounts to Include the Environ-
    ment. Washington, DC: National Academy Press.
NRC. 2005a. Decision Making for the Environment: Social and Behavioral Science Research Priori-
    ties. Washington, DC: The National Academies Press.
NRC. 2005b. Valuing Ecosystem Services: Towards Better Environmental Decision-Making. Wash-
    ington, DC: The National Academies Press.
NRC. 2007. Understanding Multiple Environmental Stresses: Report of a Workshop. Washington,
    DC: The National Academies Press.
NRC. 2008a. Water Implications of Biofuels Production in the United States. Washington, DC: The
    National Academies Press.
NRC. 2008b. Transitioning to Sustainability through Research and Development on Ecosystems
    Services and Biofuels: Workshop Summary. Washington, DC: The National Academies Press.
NRC. 2009. Science and Decisions: Advancing Risk Assessment. Washington, DC: The National
    Academies Press.
NRC. 2010. Expanding Biofuel Production: Sustainability and the Transition to Advanced Biofuels:
    Summary of a Workshop. Washington, DC: The National Academies Press.
OECD (Organisation for Economic Co-operation and Development). 2006. Cost-Benefit Analysis and
    the Environment: Recent Developments. Paris: OECD.
OECD. 2008. OECD Environmental Outlook to 2030. Paris: OECD.
OECD. 2010. Guidance on Sustainability Impact Assessment. Paris: OECD.
OEHHA (Office of Environmental Health Hazard Assessment). 2010. Cumulative Impacts: Building
    a Scientific Foundation. Office of Environmental Health Hazard Assessment, California Envi-
    ronmental Protection Agency, Sacramento, CA [online]. Available: http://oehha.ca.gov/ej/pdf/
    CIReport123110.pdf [accessed May 2, 2011].

Pearce, D. 2006. The political economy of an energy tax: The United Kingdom's Climate Change Levy. Energ. Econ. 28(2):149-158.

Pearce, D., G. Atkinson, and S. Mourato. 2006. Cost-Benefit Analysis and the Environment: Recent Developments. Paris: OECD [online]. Available: http://www.lne.be/themas/beleid/milieueconomie/downloadbare-bestanden/ME11_cost-benefit%20analysis%20and%20the%20environment%20oeso.pdf [accessed June 8, 2011].

Pereira, H.M., P.W. Leadley, V. Proença, R Alkemade, J.P. W. Scharlemann, J.F. Fernandez-Manjarrés, M.B. Araújo, P. Balvanera, R. Biggs, W.W.L. Cheung, L. Chini, H.D. Cooper, E.L. Gilman, S. Guénette, G.C. Hurtt, H.P. Huntington, G.M. Mace, T. Oberdorff, C. Revenga, P. Rodrigues, R.J. Scholes, U.R. Sumaila, and M. Walpole. 2010. Scenarios for global biodiversity in the 21st century. Science 330(6010):1496-1501.

PEW (PEW Center on Global Climate Change). 2009. Cellulosic Ethanol. Climate TechBook. PEW Center on Global Climate Change, Arlington, VA. November 2009 [online]. Available: http://www.pewclimate.org/docUploads/Cellulosic-ethanol-11-09.pdf [accessed May 2, 2011].

Polasky, S., E. Nelson, D. Pennington, and K.A. Johnson. 2011. The impact of land-use change on ecosystem services, biodiversity and returns to landowners: A case study in the State of Minnesota. Environ. Resour. Econ. 48(2):219-242.

Posner, P.L. 2004. Performance Budgeting: OMB's Performance Rating Tool Presents Opportunities and Challenges for Evaluating Program Performance: Testimony before the Subcommittee on Environment, Technology, and Standards, Committee on Science, House of Representatives, March 11, 2004. GAO-04-550T. Washington, DC: U.S. General Accounting Office [online]. Available: http://www.gao.gov/new.items/d04550t.pdf [accessed Apr. 27, 2011].

RFA (Renewable Fuels Association). 2011. Statistics. Renewable Fuels Association, Washington, DC [online]. Available: http://www.ethanolrfa.org/pages/statistics [accessed May 2, 2011].

Runge, C.F., and B. Senauer. 2007. How biofuels could starve the poor. Foreign Aff. 86(3):41-53 [online]. Available: http://www.depauw.edu/discourse/documents/How%20Biofuels%20Could%20Starve%20the%20Poor.pdf [accessed May 2, 2011].

Schwartz, P. 1991. The Art of the Long View. New York: Doubleday.

Stern, N. and C. Taylor. 2007. Climate change: Risk, ethics, and the Stern Review. Science 317(5835):203-204.

Stufflebeam, D.L. 2001. Evaluation Models. New Directions for Evaluation, No. 89. San Francisco: Jossey-Bass.

Su, J.G, R. Morello-Frosch, B.M. Jesdale, A.D. Kyle, B. Shamasunder, and M. Jerrett. 2009. An index for assessing demographic inequalities in cumulative environmental hazards with application to Los Angeles, California. Environ. Sci. Technol. 43(20):7626-7634.

Székely, F., and M. Knirsch. 2005. Responsible leadership and corporate social responsibility: Metrics for sustainable performance. Eur. Manage. J. 23(6):628-647.

UNEP (United Nations Environment Programme). 2002. Global Environmental Outlook 3. London: Earthscan.

USDA (U.S. Department of Agriculture). 2011. National Statistics for Corn. U.S. Department of Agriculture, National Agricultural Statistics Service [online]. Available: http://www.nass.usda.gov/Statistics_by_Subject/result.php?2A34ECC7-2E42-3883-BF81-9AEFA12F396E&sector=CROPS&group=FIELD%20CROPS&comm=CORN [accessed May 2, 2011].

Vedung, E. 1997. Public Policy and Program Evaluation. London: Transaction Publishers.

Wang, M., M. Wu, and H. Huo. 2007. Life-cycle energy and greenhouse gas emission impacts of different corn ethanol plant types. Environ. Res. Lett. 2(2):024001.

Zerbe, N., and T. Dedeurwaerdere. 2003. Trade, Societies and Sustainable Development SUSTRA Network: Sustainability Impact Assessment. Policy Brief Paper based on the Conclusion of the SUTRA Seminar on Sustainability Impact Assessment, March 26-27, 2003, Louvain-la-Neuve, Belgium [online]. Available: http://www.agro-montpellier.fr/sustra/publications/policy_briefs/policy-brief-sia-eng.pdf [accessed June 9, 2011].

5

How Risk Assessment and Risk Management Relate to the Sustainability Framework

As described in Chapter 4, risk assessment is an important analytic tool used to evaluate the effects of environmental stressors on ecosystem and human health. This tool has been applied over the past 25 years to facilitate management of environmental threats and remains a key analytic method in support of sustainability decision making, as envisioned by the Committee on Incorporating Sustainability in the U.S. EPA.

The formal risk assessment and risk management framework derives from a 1983 NRC report *Risk Assessment in the Federal Government: Managing the Process,* now known as the Red Book (NRC 1983). The goal was to help federal agencies, including EPA, make informed decisions about chemical agents in the setting of a growing understanding and public concern about the link between exposures to toxins and adverse effects, including cancer and birth defects. This setting of growing understanding and growing concern is similar to that faced by EPA in dealing with the challenges posed by the need for decision-making processes that fully support sustainability goals.

The NRC Red Book considered risk assessment to be an input into risk management. Building on work done in the United States and around the world, the four-step risk assessment paradigm was described as hazard identification, dose-response assessment, exposure assessment, and risk characterization. Risk management decisions were fully acknowledged to be based not only on risk assessment but also on economic, legal, and other policy-based evaluations. The risk management process involved development of regulatory options, and "evaluation of public health, economic, social, and political consequences of regulatory options." Those evaluations plus the "risk characterization," the fourth step of the risk assessment, are weighed for agency decision and actions, as illustrated in Figure 5-1.

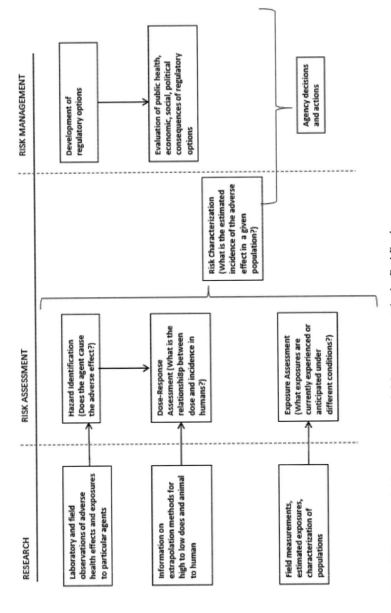

FIGURE 5-1 Elements of risk assessment and risk management in the Red Book.
SOURCE: NRC 1983.

EPA was over a decade old and had achieved many of its original objectives before the formal risk assessment paradigm was proposed in the NRC Red Book. EPA also needed the risk assessment paradigm to deal with risks that were below what was readily observable—particularly cancer risks, which were and are of great public concern (NRC 1983). EPA had already promulgated National Ambient Air Quality Standards for major air pollutants under the Clean Air Act based on non-cancer health effects. This involved direct observations of epidemiologic associations without inferring the existence of effects at air concentrations much lower than ambient levels. Risk assessment enabled effective and defensible decisions at more stringent levels corresponding to small but significant risks and was a valuable adjunct to EPA's existing "command and control" approach to regulating overt pollution of air, water, and soil.

The elements of risk assessment were already in place and being used in the United States and elsewhere, particularly for food and occupational health issues related to chemicals, before the publication of the Red Book. The NRC committee gathered ideas and did a superb job formulating a coherent approach. The Red Book committee was particularly effective in articulating the strengths and limitations of the various parts of the paradigm.

The NRC Red Book committee also had a major role in clearly defining key terms, such as "risk" and "hazard." The terms had been subject to various definitions or were used interchangeably, complicating communication and development of an agreed upon approach to risk issues. Also of note, the 1983 Red Book was not immediately adopted within EPA or elsewhere. It required several years for its general acceptance at EPA and its diffusion to state and local agencies.

INFERENCE GUIDELINES AND OPERATIONAL PROCEDURES

The Red Book (NRC 1983) stated that regulatory agencies had difficulty making decisions about a chemical agent because of the "inherent limitations," particularly uncertainties, in the science and limited analytic capacity. To bridge gaps in knowledge, the Red Book recommended that uncertainties be addressed through default inference guidelines— "an explicit statement of a predetermined choice among the options that arise in inferring human risk from data that are not fully adequate or not drawn directly from human experience" (p.51). It acknowledged the impossibility of distinguishing among such options based purely on scientific observations. These inference guidelines were seen as a way to promote consistency in analyses and avoid manipulation of outcomes. A system of guidelines, default options, and practice has developed over the years to support assessment, and a recent NRC (2009) review of EPA's risk assessment practice endorsed and emphasized the need for a clear system of defaults to support agency decision making.

LIMITATIONS OF THE RISK ASSESSMENT
AND RISK MANAGEMENT PARADIGM

EPA has had many successes in addressing the environmental problems of significant concern in the 1980s when the risk paradigm was developed. Major improvements in air and water are evident; wastes are being handled in ways less likely to pollute soil and other media; substantially lower levels of dioxins and polychlorinated biphenyls (PCBs) are present in the environment and in humans (EPA 2008).

Yet it is evident that standard risk-based regulatory approaches have limitations, including difficulty in dealing with complex problems that are not readily addressed by analyses that seek to "simplify the multidimensionality of the risk or make sense of the uncertainty" (NRC 1996) or require a volume of information and analyses that far outstrip the resources available to provide them (NRC 2006). Examples include global climate change (WHO 2003), environmental justice (NRC 1996, 2009), green chemistry (NRC 2006), nanotechnology (GAO 2010), and species loss. Recognition of the limitations in approaching these complex issues has led to approaches to widen the risk paradigm, to include the context in which the analysis is performed, the early consideration of a broad range of decision options, and the cumulative threats of multiple social, environmental, and economic stressors to health and the environment. These recent approaches can be considered to be attempts to widen the risk paradigm so as to more readily confront concerns that are central to sustainability.

Risk assessment as an analytic tool is limited in part because it works best for those chemical and physical agents that have already been emitted, in part because the nature and degree of exposure is better understood and can be monitored. In terms used in public health, risk assessment is pertinent mostly to secondary prevention (there already is a problem) rather than primary prevention (the problem never occurs) which arguably is included within the Sustainability Framework. Other issues include the inability to scientifically verify low-level risks; significant delays in risk assessment, particularly at the national level (NRC 2009); challenges to the toxicologic basis underlying risk assessment presented by agents such as nanoparticles (Tsuji et al. 2006) or endocrine disruptors (Welshons et al. 2003); and the absence of data on hazard or exposure for quantitative risk assessment (NRC 2009). Default assumptions developed or used in the Sustainability Assessment and Management approach need to be evidence-based and used according to current EPA policy.

EVOLUTION OF THE RISK ASSESSMENT AND
RISK MANAGEMENT PARADIGM

Risk assessment has been used by the agency as the main means for translating various types of biologic information about health effects of chemicals into measures of harmfulness to people. The growing understanding from biomedical and toxicologic research on how chemicals contribute to disease, coupled with better technologies and analytic tools for characterizing exposures, has resulted in continuing calls for refinement in risk assessment.

The increasing understanding of how people might differ in their responses to chemicals has led to calls for explicit treatment of human variability (NRC 1994, 2009) and vulnerable populations in risk assessment. The NRC's (1993, p. 3) finding that children are not little adults in terms of sensitivity and require consideration in risk assessment resulted in changes in practice, some through legislation. Along with genetics, preexisting health conditions also can drive individual differences in response to chemicals. Psychosocial stress may also influence risk (e.g., those exposed to violence have increased asthma in response to traffic-related air pollution). Environmental chemicals can add to endogenous exposures to agents that affect the same disease processes. All the above factors have raised concerns about some of the underlying default assumptions used in dose-response assessment and overall approaches to risk assessment (NRC 2009). There has been tension over taking these factors into account because defaults developed to address them may ultimately translate to greater stringency in risk management, and there is uncertainty over how these factors translate to risk.

Coincident with calls for a fuller incorporation of variability in risk assessment has been a push toward more sophisticated approaches toward uncertainty assessment. Cautionary notes have been given regarding the "unknown unknowns" and the limits of uncertainty analysis (NRC 1996). Approaches to such limitations and to scale the analysis to the decision at hand (NRC 1996, 2007, 2009) have been advanced.

At the same time, the scope of assessment has expanded. Multiple, related chemical exposures (NRC 2008), multiple environmental sources, psychosocial stressors in underserved communities, and other aspects of vulnerability has led to calls for cumulative risk assessment (NRC 2009). There is an explicit recognition that the current quantitative risk assessment system of defaults and techniques cannot address problems of such complexity adequately and that simpler approaches are needed (Callahan and Sexton 2007; NRC 2009). Over time, approaches to formally assess ecologic risk conditions that take a systems perspective have emerged. Finally, the great number of chemicals of potential concern is always increasing. The vast array of chemicals that are potential environmental contaminants include synthetic chemistry products, industrial chemicals, off releases from consumer products, combustion by-products, and environmental transformation by-products following chemical release—an array too vast to address by the chemical-by-chemical approach of toxicity testing in animals of each health effect of concern and then predicting human risk (NRC 2006).

Nanomaterials share many of these same characteristics. A recognition of the limited capacity to generate toxicity data to make evidentiary decisions about risk led a 2007 NRC committee to recommend initiating a 10-20 year national effort of developing new approaches for establishing the evidence base that would rely to a large extent on toxicogenomic screens. To reduce the costs and facilitate the development of new toxicity-testing approaches, interagency collaborations have been established within the United States, and the United States has also been a major supporter of, and contributor to, the OECD Chemicals Programme and its work on the "Mutual Acceptance of Data" and "Code of Good Laboratory Practices" (Ruffing 2010). As a result of recent chemical registration requirement, it is now apparent that over 100,000 chemical substances are in use in the European Union (EC 2011), a number far greater than expected, albeit some are used in low volumes. This finding provides further indication of the extremely limited capacity of the current risk-based system to deal with chemical management needs. Thus, although risk assessment provides a useful tool for looking at health effects for a circumscribed problem in a systematic way, it may not be up to the task of addressing many of the complex problems facing the agency.

The framework for risk-based decision making has been confused with risk assessment. For some sectors in the nongovernmental and business communities, the term "risk assessment" is of concern. For environmentalists the term can sometimes be code for license to pollute up to levels just below those that would be labeled a de minimis risk under traditional risk assessment methodology (Long et al. 2002). Further, the methodology may not account for all chemicals of concern to the community (NRC 1996), for environmental hazards from other facilities in the community, for community vulnerability (Morello-Frosch et al. 2011). Ultimately, risk assessment can be taken by the community as emblematic of their powerlessness (Freeman and Godsil 1993). From the business community perspective, risks just over a preset de minimis "bright line" can be characterized harmful and result in costly mitigation measures (Long et al. 2002). The identification of hazards even with vanishingly low exposures and risks can sometimes be felt as stigmatizing and a liability for the business community. In implementation of some environmental programs, a bright line can translate directly to a risk management response without consideration of other concerns (Goldstein 1993). These experiences can lead to challenges of the risk assessment and the science underlying it as a means to delay the possibility of costly actions.

Further the separation of risk assessment and risk management without consideration of the specifics of the regulatory decision to be made has resulted in protracted analyses that are not focused on the specific decision to be made. To address these issues, the NRC (2009) recently altered the risk assessment and risk management framework. As shown in Figure 5-2, risk assessment and other analytic tools are used as approaches to discriminate among potential alternatives or options for risk mitigation, identified early in the process.

The risk assessment and risk management paradigm forms the basis for risk-based decision making within EPA programs, such as the legacy cleanup programs Comprehensive Environmental Response, Compensation, and Liability Act (CERCLA) and Resource Conservation and Recovery Act (RCRA) and in standard setting programs, for example, NAAQS and maximum contaminant

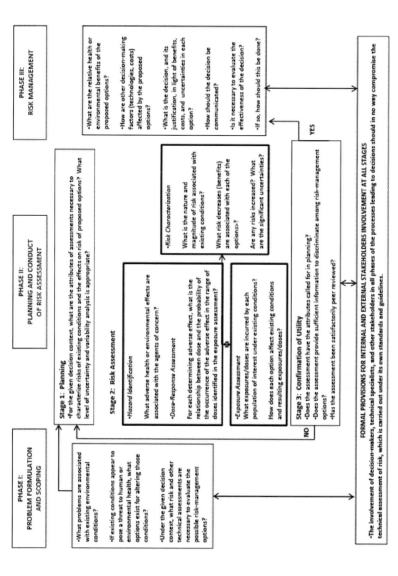

FIGURE 5-2 Framework for risk-based decision making.
SOURCE: NRC 2009

levels (MCLs) set under the Safe Drinking Water Act. In these programs, factors other than risk are also considered and are consistent with the risk management framework in the Red Book (NRC 1983) and *Science and Decisions* (NRC 2009). However, with risk reduction as the primary objective, full consideration has generally not been given to all three sustainability pillars.

Here are two examples showing how factors other than risk are taken into account in risk management—one from setting drinking-water standards and the other from the legacy cleanup programs. In these programs, EPA often defines an ideal goal. An analogue in the legacy cleanup programs is the setting of preliminary remediation goals (PRGs)—starting points for setting cleanup levels in soil. In this situation, the risk management decision may involve an assessment of the technical feasibility of achieving a given goal. In comparing alternative risk management options in the legacy cleanup programs, numerous factors can be considered under the EPA National Contingency Plan. More explicit consideration of sustainability factors is now being considered by EPA under its green remediation strategy and is applied in some cases. Efforts have been accelerated to perform cleanups in the most environmentally sustainable manner (e.g., renewable energy use, minimum waste generation, and reduction of life-cycle greenhouse gas emissions) and to address environmental justice inequities. Analyses to characterize risks to the community, costs, community improvements, and trade-offs can provide the basis for choosing practicable management options that mitigate significant risks while providing benefits for the surrounding community, and meeting other sustainability goals. Thus, elements of sustainability assessment and thinking are increasingly being incorporated into environmental remediation decision making.

As required by the Safe Drinking Water Act, a maximum contaminant level goal (MCLG) of zero is specified for chemicals suspected of being carcinogens and having linear dose response relationships. Zero is clearly an unattainable goal. The actual enforceable standard, however, is based on a small number of technical factors, such as analytic detection limits and cost, resulting in a regulatory standard—a Maximum Contaminant Level (MCL)—that is greater than zero. For example, the MCLG for trichloroethylene is zero, but the MCL is currently 5 ppb.

THE INTERFACE BETWEEN THE RISK ASSESSMENT AND RISK MANAGEMENT PARADIGM AND SUSTAINABILITY

The committee's Statement of Task includes the question, "How can the EPA decision-making process rooted in the risk assessment/risk management paradigm be integrated into this new sustainability framework?" To respond to this charge, the committee has separated risk assessment from risk management. The four-step risk assessment paradigm will continue to be valuable in identifying and managing risks, quantitatively or qualitatively. The conceptual approach to identifying an intrinsic hazard, understanding the link between the hazard and an unwanted effect, calculating the extent to which humans or ecosystems are exposed to that hazard, and characterizing the resultant risk in a manner pertinent to policy makers and the public can be extended beyond the risk of chemical and physical agents in the environment. It can also incorporate qualitative approaches and other approaches to express risk or cumulative risks to address a wider range of issues, but tools will be needed to make that a reality. Accordingly, the committee reaffirms the value of risk assessment, finds it to be a useful tool for sustainability, and encourages the further development of risk assessment tools, such as to address cumulative risks, to improve its usefulness.

The committee notes that the term "risk management" is used in two ways: as a formal description of EPA's policies related to control of environmental risks and as an informal term denoting any EPA approach to management of current or potential threats. Sustainability goes beyond risk management in being primarily concerned with maximizing benefit, while addressing risks of concern, rather than being an exercise focused mainly on achieving de minimis risk. The focus on de minimis risk sometimes includes risk-risk and risk-benefit trade-offs in management decisions, but does not necessarily or typically encompass the social (including health), environmental, and economic pillars of sustainability. Risk management in either the formal or informal use of the term does not fully encompass the sustainability paradigm in which the management of risk is perceived as an opportunity to maximize benefits while controlling environmental harm. Table 5-1 presents the committee's comparison of key features of risk assessment/risk management framework with sustainability.

The Sustainability Framework can include each of the basic elements of the Red Book and recent (NRC 2009) risk assessment and risk management paradigms. In some cases, however, a formal four-step risk assessment will not help to discriminate among potential decision options and should not be performed. For example, the time frame for the decision may not permit the type of data gathering needed to support risk assessment, or the nature of the problem is such that a risk assessment would be noninformative. For decision processes in which four-step risk assessment is included, the Sustainability Framework can be viewed as representing the risk paradigm expanded and adapted to address sustainability goals, as illustrated in Figure 5-3, where the first two steps are dealt with in Phase I.

Under the Sustainability Framework, in addition to considering possible technology options to minimize risk, consideration of opportunities and options for improvements along social, environmental, and economic dimensions would also be elements corresponding to the risk assessment and risk management (RA/RM) framework's phase I, the problem formulation and scoping phase. Phase I would also include identifying possible state, local, and other federal collaborators that may participate in the project. Stakeholders—interested parties affected by the decision—could help identify options.

TABLE 5-1 Differences Between Features of Risk Assessment and Risk Management and Sustainable Development

Feature	Risk Assessment and Risk Management	Sustainability
Relation to EPA statutes	Typically required	An exercise of discretion in implementation of statutes
Driver	Statutes and implementing regulations; need to defend ultimate decision publicly and in court	Opportunity to reduce costs; increase social, environmental, and economic benefits while meeting statutory requirements to mitigate risk
Questions to be answered	What is the risk? What action to take in face of risk? At times, what is the result of cost–benefit analysis?	How little harm is possible? How can we maximize social, environmental, and economic benefits?
Number of decisions to be made at one time	Typically fewer	Could be coordinated suite of decisions affecting, e.g., a particular place
Number of agencies involved	Less	More
Subject matter	Pollutants and chemicals	Social, environmental, and economic factors (including not only pollutants and chemicals but also, e.g., community, jobs, and quality of life)
Focal points	Ecologic and health risks from chemicals	Social (including but not limited to public health), environmental, and economic impacts
Metrics	Typically quantitative (for human health)	Quantitative and qualitative
Formal assessment process required?	Typically yes	No, but formal processes can be used
Who does it in EPA?	Primarily risk assessors (e.g., toxicologists, epidemiologists, exposure assessors and sometimes economists); program risk managers; limited collaboration with external agencies	Multidisciplinary and potentially multi-program teams; more collaboration with outside federal, state, and local agencies
Stakeholders involved?	Beyond scientific peer review and formal public notice and comment requirements, depends on program	Generally more inclusive and broader due to the questions
Nature of stakeholder involvement?	Often part of routine public comment and review	Discussion at many different levels; more "outside the box"
Relationship among stakeholders	Often adversarial	Potentially more collaborative

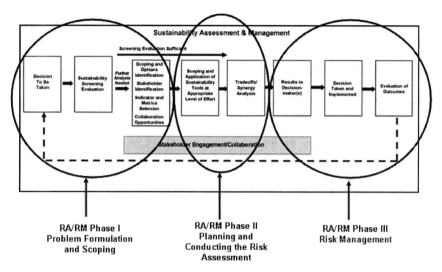

FIGURE 5-3 Correspondence between sustainability assessment and management elements and risk assessment and risk management (RA/RM) framework.

Corresponding to an RA/RM phase II under the Sustainability Framework in addition to planning and conducting the risk assessment, other assessments would be included to help discriminate among policy options. Analysis would address critical social, environmental, and economic features associated with the different options. Analyses of trade-offs between different features associated with the different policy options would also be conducted. All analyses would be sized in terms of intensity and scope to provide outcomes that enable selection across options. It is in this phase that analyses would, if required or desirable, receive technical peer review and stakeholder comment.

Corresponding to an RA/RM phase III under the Sustainability Framework, the decision makers would deliberate on the results of these assessments, struggle with trade-offs, and make decisions. The implementation of the decision would be followed by an evaluation of its effectiveness.

FINDINGS AND RECOMMENDATIONS

5.1. Key Finding: The risk assessment and risk management frameworks articulated by the NRC (1983-2009) are analogous with the committee's proposed Sustainability Assessment and Management approach. However, differences in their overall goals how greater complexity in the Sustainability Framework components of scoping, analysis, deliberation, and decision making. The four-step risk assessment process, as envisioned by the NRC

(1983) Red Book, is an important component and tool used to inform decisions in the Sustainability Assessment and Management approach (p.86-89).

5.1. Key Recommendation: EPA should include risk assessment as a tool, when appropriate, as a key input into its sustainability decision making.

REFERENCES

Callahan, M.A., and K. Sexton. 2007. If cumulative risk assessment is the answer, what is the question. Environ. Health Perspect. 115(5):799-806.

EC (European Commission). 2011. REACH. European Commission [online]. Available: http://ec.europa.eu/environment/chemicals/reach/reach_intro.htm [accessed May 31, 2011].

EPA (U.S. Environmental Protection Agency). 2008. EPA's Report on the Environment. EPA/600/R-07/045F. U.S. Environmental Protection Agency, Washington, DC. May 2008 [online]. Available: http://www.epa.gov/ncea/roe/docs/roe_final/EPAROE_FINAL_2008.PDF [accessed June 8, 2011].

Freeman, J.S., and R.D. Godsil. 1993. The question of risk: Incorporating community perceptions into environmental risk assessments. Fordham Urban L.J. 21(3):547-576.

GAO (U.S. Government Accountability Office). 2010. Nanotechnology: Nanomaterials Are Widely Used in Commerce but EPA Faces Challenges Regulating Risk. GAO 10-549, May 2010 [online]. Available: http://www.gao.gov/new.items/d10549.pdf [accessed June 8, 2011].

Goldstein, B.D. 1993. If risk management is broke, why fix risk assessment? EPA J. Jan:37-38.

Long, T.F., M.L. Gargas, R.P. Hubner, and R.G. Tardiff. 2002. The role of risk assessment in redeveloping brownfields sites. Pp. 281-326 in Brownfields: A Comprehensive Guide to Redeveloping Contaminated Property, 2nd Ed., T Davies, ed. Chicago, IL: Section of the Environment, Energy, and Resources, American Bar Association.

Morello-Frosch, R., M. Zuk, M. Jerrett, B. Shamasunder, and A.D. Kyle. 2011. Understanding the cumulative impacts of inequities in environmental health: Implications for policy. Health Aff. 30(5):879-887.

NRC (National Research Council). 1983. Risk Assessment in the Federal Government: Managing the Process. Washington, DC: National Academy Press.

NRC. 1993. Pesticides in the Diets of Infants and Children. Washington, DC: National Academy Press.

NRC. 1994. Science and Judgment in Risk Assessment. Washington, DC: National Academy Press.

NRC. 1996. Understanding Risk: Informing Decisions in a Democratic Society. Washington, DC: National Academy Press.

NRC. 2000. Ecological Indicators for the Nation. Washington, DC: National Academy Press.

NRC. 2006. Toxicity Testing for Assessment of Environmental Agents: Interim Report. Washington, DC: The National Academies Press.

NRC. 2007. Toxicity Testing in the 21st Century: A Vision and a Strategy. Washington, DC: The National Academies Press.

NRC. 2008. Phthalates and Cumulative Risk Assessment: The Task Ahead. Washington, DC: The National Academies Press.

NRC. 2009. Science and Decisions: Advancing Risk Assessment. Washington, DC: The National Academies Press.

Ruffing, K.G. 2010. The role of the Organization for Economic Cooperation and Development in environmental policy making. Rev. Environ. Econ. Policy 4(2):199-220.

Tsuji, J.S., A.D. Maynard, P.C. Howard, J.T. James, C.W. Lam, D.B. Warheit, and A.B. Santamaria. 2006. Research strategies for safety evaluation of nanomaterials, Part IV: Risk assessment of nanoparticles. Toxicol. Sci. 89(1):42-50.

Welshons, W.V., K.A. Thayer, B.M. Judy, J.A. Taylor, E.M. Curran, and F.S. vom Saal. 2003. Large effects from small exposures. I. Mechanisms for endocrine-disrupting chemicals with estrogenic activity. Environ. Health Perspect. 111(8):994-1006.

WHO (World Health Organization). 2003. Climate Change and Human Health: Risks and Responses, A.J. McMichael, D.H. Campbell-Lendrum, C.F. Corvalan, K.L. Ebi, A.K. Githeko, J.D. Scheraga, and A. Woodward, eds. Geneva: World Health Organization [online]. Available: http://www.who.int/globalchange/publications/climchange.pdf [accessed June 8, 2011].

6

Changing the Culture in EPA

EFFECTING CULTURAL CHANGE IN THE AGENCY

Implementation of sustainability efforts in EPA will be fostered if the culture of the agency changes so that sustainability is a common way of thinking for everyone in the agency. This would provide agency personnel with the opportunity to integrate sustainability into their work. Although EPA has developed significant tools and knowledge with which to implement the Sustainability Framework, further innovations and implementing transitions (long-term changes in an encompassing system that serves a basic societal function) will be necessary to meet the increasingly complex challenges it faces (Elzen and Wieczorek 2005). Therefore, EPA should foster cultural change and innovation at all levels of the organization to meet the challenges of increasingly complex problems. EPA also would benefit from the experience of other organizations and countries that are considering similar cultural changes concerning sustainability and from a growing literature on effecting this change (Kemp et al. 2007, Nill and Kemp 2009).

The agency has had several successful initiatives to manage significant internal change. The early efforts to educate staff and appointees on risk assessment and its use in decision making is one example. Others include a subsequent effort to change how risk characterization was performed as well as implementing pollution prevention activities. Similar efforts should be undertaken regarding sustainability as a first step in its wider use in the agency. Materials developed for this effort should be shared broadly outside the agency for the benefit of its stakeholders and its partners in program implementation.

To be responsive to a sustainability vision, there needs to be a broadening of disciplinary approaches toward understanding underlying processes. Tools that are pertinent to making sustainability decisions need to be developed. Indicator sets and appropriate metrics need to be made available to program and regional offices that assess programs; that project progress toward sustainability; and that permit the public and decision makers to address sustainability issues efficiently and effectively (Box 6-1). Specifically, EPA must recognize the challenges of the long-term impacts of environmental management decisions made in the near-term time frame.

Creating a Culture of Sustainability

Incorporating sustainability into EPA's mode of operation will require a shift toward a more systems-based approach that integrates multiple media, with multiple objectives in social, environmental, and economic pillars and considers both short-term and long-term consequences. Changing thinking within the organization is a large task with responsibilities throughout the agency. It requires a clear statement of principle about the importance of sustainability for the agency from

---

BOX 6-1
Everglades Restoration:
The Comprehensive Everglades Restoration Project

Damage to the Everglades due to the lack of a sustainable approach to land use in southern Florida led to a congressional authorization of $13.5 billion to fund the Comprehensive Everglades Restoration Project (CERP) in 2000. Half of this funding was expected to be raised by multiple local sponsors. Over the past 5 years, nearly $2 billion in combined contributions from federal and state partners has been provided to support CERP projects; the federal government expended almost $259 million, and it is estimated that local sponsors spent approximately $270 million on activities not related to land acquisition, which remains a major expense. As of 2009, the State of Florida has spent $1.26 billion to purchase more than 230,000 acres (CERP 2010). The restoration is expected to take a total of 30 years, but when completed, it is hoped that the hydrologic characteristics will be restored to previous levels to serve the natural and human demands on the ecosystem (State of Florida 2011). The Florida Everglades is a large and diverse aquatic ecosystem that over many years has been dramatically altered to increase regional productivity through flood control measures, water-supply needs, and agricultural production. The NRC Third Biennial Review of CERP (NRC 2011) recommended that the South Florida Water Management District (SFWMD) "complete a comprehensive scientific, technical, and cost-effectiveness analysis as a basis for assessing potential short- and long-term restoration alternatives and for optimizing restoration outcomes given state and federal financial constraints" (p. 12). The Everglades restoration provides another example where a sustainability framework that integrates the social, environmental, and economic dimensions required by such an analysis could potentially reduce the cost of the current multibillion dollar effort.

---

the top. It also requires bottom-up incorporation of sustainability into everyday operation for regions and programs. In addition to changes in thinking, incorporating sustainability into EPA operations will require integrating and extending existing approaches and, in some cases, developing new approaches. The agency can accelerate its cultural change by practicing sustainability in its day-to-day operations.

Including progress in programs and projects that incorporate the sustainability approach in individual performance reviews will further accelerate the cultural change, as will hiring and training staff with expertise in sustainability issues (McKenzie-Mohr and Smith 1999). Communicating the basic elements of sustainability is an important step in achieving the cultural change within the agency. EPA would benefit from undertaking an employee communication and education program to effect the cultural change necessary to support the adoption of the Sustainability Framework and to enable employees at all levels to participate in its implementation.

EPA would also benefit from hiring multidisciplinary professionals who are proficient in many disciplines. The Science Advisory Board (SAB) in 2010 recommended that EPA increase its investment in social, behavioral, and decision sciences across the Office of Research and Development (ORD) research portfolio to help EPA develop and implement workable solutions to major environmental challenges (EPASAB 2010). EPA would also benefit from the inclusion of multiple fields, such as integrated transport and landscape architecture and ecosystem services, that relate to managing landscapes in an ecologic way and for conservation purposes (Lindenmayer et al. 2008). At first, the strategy may be to hire from outside sectors leaders to aid EPA in shifting to a more cross-cutting mind-set (NRC 2010). The ORD has focused on six broad research themes in their approach to achieving sustainability (EPA 2007), and EPA could further collaboration among existing professional expertise to encourage dialogue and understanding of the various fields and work already within EPA.

In addition to being used by staff at all levels, the Sustainability Framework could also be incorporated into all agency advisory committees' considerations and recommendations. An internal advisory group focused on sustainability during the initial phases of effecting cultural change and adopting the Sustainability Framework could be useful to the agency. This group could be part of the already functioning communication between the 10 regional administrators. They have on-the-ground knowledge of, and in many cases, connections to local leaders, who may be able to understand and advise on sustainability initiatives that could have the greatest impacts if leveraged and supported by the agency.

The Committee on Incorporating Sustainability in the U.S. EPA recognizes that further reorganization within EPA may be the consequence of the further pursuit of sustainability. However, the incorporation of a culture of sustainability within the operations of the agency is essential, and managing this change should take precedence over organizational change.

A final and important point to make about instigating cultural change within the agency deals with putting money where the focus is. To change the direction of an organization, its budgets must reflect that change, and, to that end, modifying budget proposals to reflect the sustainability objectives at the regional and program levels might help. It might also be helpful to make room within the agency's budget for national and regional work to test and implement sustainability initiatives. Ensuring that available federal funding more expressly addresses sustainability's social, environmental, and economic objectives is likely to produce greater benefits for each dollar expended. The committee recognizes that, during tight budget times, obtaining new funding will be difficult and will require creative and continual evaluation of existing funding initiatives on a program-by-program basis. Funding partnership opportunities sought out at every juncture will help to ensure continuity, as leveraging funding successfully benefits a variety of stakeholders. A 2009 NRC workshop report on enhancing the effectiveness of sustainability partnerships (NRC 2009) provides guidance on this issue.

RESEARCH AND DEVELOPMENT

Given the overall mission of EPA's ORD,[1] it can play a lead role in the development of specific tools (Chapter 4) that are useful for implementing the Sustainability Framework (Chapter 3). Additionally, ORD can identify relevant research needs to address many of the unresolved challenges. Many of the issues discussed in this report highlight the need to develop scientific and analytic tools to support this framework. The committee provides here suggestions on conceptual approaches that could form the bases for establishing research and development (R&D) priorities and incorporating sustainability concepts into this decision process.

Incorporating sustainability into EPA operations does not require fundamentally new approaches; in many instances, simply modifying or expanding on the application of existing tools so that they are more integrative and intergenerational will be sufficient. On the other hand, attention to some issues raised by the Sustainability Framework will depend upon R&D support. Recognition of the need for EPA's science to be responsive to the increasing complexity of EPA's challenges has been voiced in many external and internal reviews (Powell 1999). Despite this recognition and the improved understanding and development of tools that fit well under a sustainability paradigm, EPA must go further to recognize sustainability specifically as a key driver for its research program.

---

[1] The mission of the ORD is to "perform research and development to identify, understand, and solve current and future environmental problems; provide responsive technical support to EPA's mission; integrate the work of ORD's scientific partners (other agencies, nations, private sector organizations, and academia); and provide leadership in addressing emerging environmental issues and in advancing the science and technology of risk assessment and risk management" (EPA 2010a).

That the environmental science community has come a long way but needs to go further is exemplified by the reports of two companion NRC committees, which were funded by Congress and EPA to explore the role and the future direction of EPA's science and technology (NRC 1997, 2000a). These companion documents emphasized the importance of ORD in performing the long-term core research necessary to achieve EPA's mission. As noted in the latter report,

> The very nature of the problems faced by EPA has been changing dramatically, and surprises have become common. EPA was created in 1970 with the limited understanding of environmental issues available at that time, including some concepts that are now largely outdated and rapidly being subsumed *in new concepts such as sustainable development and industrial ecology* (EPASAB 1988; NAE 1994; NAPA 1994; OSTP 1994). These concepts envision the integration of environmental science and technology throughout the entire economy. They are not simply (or in many cases even primarily) concerned with reducing existing impacts or ensuring compliance with so-called "end-of-pipe" regulations. If ORD is to participate effectively in developing and implementing new concepts and policy directions, its scope of activities should be appropriately expansive. ORD should address not only the individual pollution-related problems that have traditionally concerned EPA, but also the research on complex topics such as *sustainable development* and biological diversity [italics added] (NRC 2000a, p.43).

The research activities considered in these two documents fit primarily within the goal of decreasing pollution rather than maximizing overall social, environmental, and economic benefits. The 1997 committee's formulation of four components of environmental research—understanding underlying processes, developing tools, acquiring data, and communicating results—remains central to the R&D necessary to achieve EPA's sustainability goals. The above paragraph is one of the few instances in either NRC report that sustainable development or sustainability is mentioned.

The ORD has already taken a number of steps along this path. Hecht (2009) noted that EPA launched its *Sustainability Research Strategy* "with the dual goals of advancing an understanding of biological, physical, and chemical interactions through a systems and life-cycle approach, and developing effective models, tools and metrics that enable decision makers to achieve sustainable outcomes." In this strategy, ORD laid out a strategy for sustainability research at EPA that included important elements:

> Our *Sustainability Research Strategy* rests on the recognition that sustainable environmental outcomes must be achieved in a systems-based and multimedia context that focuses on the environment without neglecting the roles of economic patterns and human behavior. This recognition begets a fundamental change in research design. In a systems-based approach, the traditional goals of achieving clean air or water or protecting ecosystems and human health can be

fully understood only through a multimedia approach. EPA and its partners will develop integrating decision support tools (models, methodologies, and technologies) and supporting data and analysis that will guide decision makers toward environmental sustainability and sustainable development. (EPA 2007, p.6)

The committee did not have the time or detailed information necessary to develop specific recommendations on changes or additions to the current R&D programs within the ORD. The committee notes, however, the six major research themes that the ORD identified emphasize an integrated and systems-based approach to achieve sustainability (EPA 2007):

- **Renewable Resource Systems:** determining how best to obtain the benefits that renewable resources provide, while considering the system-wide effects their use has on the regenerative capacity of the entire system
- **Non-Renewable Resource Systems:** promoting sustainable management of non-renewable resource operations and supporting the shift to renewable resources
- **Long-Term Chemical and Biological Impacts:** assessing and eliminating the long-term impacts posed by harmful chemical and biological materials in order to improve our use of materials, shift to environmentally preferable materials, and protect human health. Developing alternate chemicals, new industrial processes, and new decision support tools for evaluating environmental dimensions of new chemicals and processes
- **Human-Built Systems and Land Use:** researching sustainable building design and efficiency, management of urban systems, life cycle assessment for building design and land use, and decision support tools for urban land development and revitalization
- **Economics and Human Behavior:** developing ecosystem valuation methods and analyzing the role of incentives in decision making and the causes of market failures
- **Information and Decision Making:** identifying appropriate indicators and ensuring their quality

If sustainability is to become the "true North" or main vision for ORD, clearly the R&D programs and projects would need to reflect this new vision. In this context, there are key themes for ORD to consider in achieving alignment of R&D activities within the Sustainability Framework. Topical areas for ORD to consider in modification of current research priorities and in establishing future ones include the following:

- Create a suite of decision-support tools for long-term impact analysis and simple decision tools for use by communities.

- Set research priorities using strategic objectives related in part to sustainability.
- Develop system models capable of providing projections and develop alternative projections for present and future outcomes for key types of issues.
- Develop robust methods that can readily incorporate uncertainty, variability, vulnerability, and resilience.

In July 2010, the SAB commented on the ORD's strategic research directions, identifying several priority issues, including the importance of a systems approach and integrated transdisciplinary research (ITR), the importance of incorporating social and behavioral sciences into ORD's research programs, and the need for the agency to provide leadership in establishing multiagency partnership for leveraging of resources (Swackhamer 2010). In response, Lisa Jackson, the EPA administrator, agreed with each of the SAB's points. Regarding ITR, Ms. Jackson stated that the agency believes that "ITR can potentially be a driving force in the EPA's pursuit of sustainable environmental solutions" and reiterated the agency's focus on effectively implementing ITR through research partnerships and in adopting a systems approach to research planning (Jackson 2010).

Incorporating Sustainability Through Place- and Programmatic-Based Opportunities for Collaboration

Incorporating sustainability requires a systematic, programmatic approach: Adopt a framework, retrain and educate employees, alter fiscal priorities to reflect the framework, and evaluate on-the-ground initiatives for effectiveness. Place-based projects, or projects that are based in a specific locale with measurable outcomes, whether in collaboration with others or under the sole auspices of EPA, are a logical step in creating the culture change within the agency, as recommended by the Sustainability Framework. Planning and executing successful place-based projects will assist the agency's programmatic approach to cultural change. An early pilot effort in the use of sustainability in a well-defined programmatic area will help facilitate change in EPA's headquarters. Successfully incorporating sustainability into any organization requires commitment from both the top down and the bottom up as both a directive and an owned initiative. Therefore, EPA should consider a set of place-based and programmatic-based pilots to develop the cultural change necessary for the successful adoption of the Sustainability Framework; however, place-based projects have the advantage that they are more often collaborations both within and outside EPA.

During committee discussions with agency stakeholders, a number of examples of the incorporation of sustainability in carrying out place-based and programmatic-based projects were given. Examples include the following:

- Boston's Fairmont Rail Corridor was redeveloped (Box 6-2).
- The city of New York invested in protecting and restoring its watershed to meet EPA's requirement to filter drinking water from surface-water sources to protect the public from waterborne diseases (Box 6-3).
- The University of Minnesota Water Resources Center was tasked by the state legislature to develop a framework that addresses aspects of water sustainability in that state. These include drinking water, stormwater, agricultural and industrial water use, surface and groundwater interactions, infrastructure water needs, and water use within the context of predicted changes in climate, demographics, and land use. The resulting framework identified the 10 priority issues that present the challenges and the solutions to address those challenges. Each of these priority

---

BOX 6-2
Redevelopment of Boston's Fairmont Rail Corridor:
Addressing Environmental Justice Issues Through
Multi-Agency and Community Collaboration

The redevelopment of Boston's Fairmont Rail Corridor provides an excellent example of the benefits of multiagency and community collaboration, along with EPA's contribution of technical assistance, some funding, and its deep expertise in requiring and overseeing industrial site cleanups across the region and the country. In 2009, EPA, the U.S. Department of Housing and Urban Development (HUD), and the U.S. Department of Transportation announced a new initiative, the Partnership for Sustainable Communities, designed to improve agency collaboration in assisting communities to become more sustainable. The agencies selected five pilot communities, including Boston's Fairmont Rail Corridor, where there were multiple brownfield sites, along with a need for more affordable housing and better public transit options for renovation. Boston's Fairmont Rail Corridor, a commuter rail line, passes through some of the city's poorest neighborhoods without stopping. The area immediately surrounding the rail line is economically depressed despite the area's proximity to downtown Boston. To catalyze development in these neighborhoods, four new stations are being built and two are being renovated. EPA will clean up more than 30 brownfield sites within a half-mile of the new and renovated stations to provide room for potential development. EPA has also committed to providing technical assistance to a "Green Jobs Incubator" on one brownfield site. HUD is assisting in the development of 2,000 new housing units along the corridor, including some affordable housing for current residents. In addition, the housing agencies are coordinating with the city, several community development corporations, the Boston Foundation, and others to create jobs and encourage more development of affordable housing. Two stations have been rehabilitated and several neighborhood bridges have been de-leaded and repainted. The new stations, which are expected to expand development opportunities, are set to be completed in 2012 (EPA 2010b).

BOX 6-3
Approving New York City's Water Supply Protection System

In 1997, the City of New York avoided a $9 billion investment in filtration plants by making a $1.5 billion investment in protecting and restoring the watershed (UNEP 2007). New York obtained a waiver from EPA requirement to filter drinking water from surface-water sources to protect the public from waterborne diseases. This agreement was founded on the city's long-standing water-supply protection strategy for protecting forestland and building reservoirs and aqueducts in the Croton, Catskill, and Delaware catchments (UNEP 2007). The strategy was reviewed in 2000 by the NRC, which made recommendations on water-quality monitoring, and overall, found that "the concept of balancing watershed rules and regulations with targeted support of watershed community development is a reasonable strategy for New York City and possibly other water supplies" (NRC 2000b, p. 2). The system today is reliant on a 5,200-square-kilometer catchment system that spans eight counties. The agreement, extended for another 10 years in 2009, provides for community management and protection of the watershed, septic system repair and replacement, riparian and buffer protection, and continued acquisition of forestland and easements (UNEP 2007). The city's history of using this approach provided some evidence of its potential for the future. There are major economic and social benefits of this unconventional strategy, such as the long-term provision of recreational resources as well as health benefits and other amenities of great value for the communities in the watershed.

---

issues falls within the three areas that define sustainability: social, environmental, and economic (UMN-WRC 2011).

Pursuing sustainability-related projects where collaboration is fostered with regions, states, communities, nongovernmental organizations, businesses, tribes, and internationally allows the partners to learn from each other and promote the dissemination of best practices. Experiences from these collaborations and place- and programmatic-based efforts will provide engineers and social, physical, and biologic scientists with new challenges for developing tools, indicators, and more fundamental research. For example, EPA has demonstrated that it can support sustainable land-use activities at the state and local level without being overly prescriptive (Box 6-4). Under the committee's framework, EPA could consider how best to work with state and local health departments to complement and reinforce their land-use guidance.

A continued special effort to collaborate with other federal agencies who share responsibility at the federal level for sustainability is called for. Collaborations such as the Federal partnership for sustainable communities between the U.S. Department of Transportation, the U.S. Department of Housing and Urban Development, and EPA promise similar advances at the federal level that are

---

BOX 6-4
An Example of EPA's Role in Facilitating State Activities
That Achieve Environmental Goals:
Improving Air Quality Through Land-Use Planning

EPA often provides guidance without specifying the compliance approach to states on the best ways to meet regulatory requirements. For example, EPA's Transportation Air Quality Center has published *EPA Guidance: Improving Air Quality Through Land Use Activities* (EPA 2001). It includes EPA's recommendations on how land-use planning can be part of a state implementation plan (SIP) to meet air-quality requirements related to mobile sources. Most important, it describes the modeling and accounting processes that states can use to get credit under EPA's SIP approval process. Guidance is also given about land-use planning that would help the state in the conformity process required by the Clean Air Act to assess the compliance of transportation planning with air-quality goals. EPA's document includes examples of credit given to the state for voluntary planning by local individuals—for example, a housing development that was planned in a way to decrease miles driven while improving recreational values and decreasing water use. Although not requiring specific land-use activities, EPA provides guidance and assessment techniques that encourage achieving air quality goals. Note that under the sustainability framework, EPA might join with public health agencies in providing complementary guidance; for example, a decrease in vehicle miles traveled is also pertinent to the issue of obesity.

---

being realized in other collaborations. It is important for EPA to make a conscious effort to identify stakeholders for individual initiatives and plan strategically during each budget year to fund efforts and to incorporate the outcomes of those efforts to agency performance measures. Exploring fiscal and technical partnerships with other federal agencies and at the regional and local levels in all sectors will increase sustainable leveraging potential (see Box 6-5 for an example of agency collaboration to promote more sustainable approaches to roof design). Sustainability cannot be achieved at a national level through EPA's efforts alone, so placing emphasis on making decisions and funding efforts in conjunction with other agencies and partners, especially at the federal level, is key. Empowering interagency and regional collaborations will foster accelerated progress and successes in achieving sustainability objectives and will support other agencies' efforts to adopt sustainability in their activities.

Promoting partnerships when developing approaches to public problems will ensure that EPA is both a voice at the table and a driver for full and careful considerations of decisions that have an impact on the environment (subsidy and development of alternative fuels, e.g.). It is also important to promote a diverse mix of governmental partners and stakeholders at the local, regional, and national levels in community and place-based decision making, which will allow the

---

BOX 6-5
Growing Collaboration on Redesigning Roofs

The U.S, Department of Energy and EPA along with state and local agencies are providing guidance and support for more sustainable approaches to roof design (DOE/EPA 2011). Covering a roof with vegetation lowers energy costs for the building, while conveying a wide range of environmental benefits. These include a decrease in the urban-heat island, and lower emissions of greenhouse gases and other air pollutants; the buffering of storm-water runoff; and the provision of desirable habitat for humans and for many species. Cool-roof technology consists of covering roofs with reflecting substances that have been documented by DOE-supported studies to result in lower energy costs over the lifetime of the building (DOE 2010). Scientists from the University of California at Los Angeles with public and private sponsorship, recently estimated that there is more than sufficient space on household roofs in Los Angeles for solar cells to power the entire city (DeShazo et al. 2011). A project in Philadelphia, the "Coolest Block" contest brought local communities together to compete for energy-efficiency improvements in their homes. The winning community was provided with energy-efficient technology—application of elastomeric roof coatings. This effort is in line with Philadelphia Mayor Michael A. Nutter's recent legislation requiring all new residential and commercial construction in the city to include reflective or green roofing (City of Philadelphia 2011).

---

agency to maximize every opportunity to collaborate with and leverage resources from all sectors to promote a cultural change that will support sustainability as a mind-set.

External Cultural Change

Sustainability can be promoted by EPA in its public information and education programs through technical assistance, incentives, and regulatory programs to local governments and other stakeholders. A national education campaign to clarify what sustainability is and how it changes the way the nation operates would be a beneficial starting point. The campaign could be rolled out to address the widest possible audience so that Americans could understand the importance of considering the environment, the economy, and society in all decisions. Making sustainability information accessible and available to the public needs to be a priority (Bruch et al. 2009).

There is also a very real need for education and sharing of national and international best-management practices for sustainability, especially at a local level. This sharing of practices can be done by posting examples online, and through stakeholder webinars, workshops, and field collaboration. It is especially necessary to engage in this type of community education when new permits are introduced. Not only do local department heads need this training and information, but local designers and contractors do as well so that the design process can be informed and collaborative in nature, resulting in the reduction of perceived risk that is typically incorporated into project design and construction fees when there is discomfort with the concepts.

Technical assistance should be available to stakeholders. Nationally, EPA would become more available and interactive with the public. The flow of question and answer will benefit both the agency in knowing what needs funding or support and the stakeholder, who to date does not have a comprehensive source of knowledge available to them about the overarching concepts of sustainability. This type of dialogue will also help EPA to evaluate how it can change to aid a sustainable culture change on a broader level.

EPA could also contribute to an effort in conjunction with public and private stakeholders to develop, post, and maintain a Wikipedia analogue for sustainability practices, laws, and policies at the local, state, national, and even international level. The purpose of this web-based program would be to provide up-to-date and high-quality information about what practices and processes work most effectively for sustainability.

In addition to public education and information, incentives could be created for sustainable behavior and disincentives for unsustainable behavior. Some of the best positive motivators include incentives that involve fiscal contributions from the agency to drive behavior change and public recognition of innovative achievements. Nonfinancial incentives may be the best pursuit in times of economic difficulties for the agency. Continuing sustainable best-practice and innovation awards that are high profile and well publicized is a positive way to reinforce the good work already in progress in communities and to establish healthy competition for recognition, which can grow local economies through marketing and public perception. These awards are strictly honorific in nature but can be highly publicized on a national level. For example, EPA is already providing this type of recognition in its green chemistry and brownfield programs.

Another consideration for the agency is an adoption of a "green star program," which would be an offshoot of the green products for businesses but be geared toward the civilian consumers (see Box 6-6). A program of this type would include reporting to the consumer a broader range of product impacts, much like Wal-Mart's new carbon footprint index. Programs such as Energy Star are great examples of EPA's current collaboration with other government agencies and active interaction with stakeholders at all levels.

Consideration should be given to evaluation of fast tracking and slow tracking sustainability-related regulations. An example is issuing new stormwater permits in a timely manner and offering technical assistance to assure a community that new initiatives, such as bioswales and permeable pavement, will be counted as credits toward meeting targets. Without this time and technical investment from EPA to help communities trust new ways of managing runoff, there is a tendency

BOX 6-6
Presidential Green Chemistry Awards

The concept of green chemistry was expounded by Paul Anastas and John Warner in their 2000 seminal book, *Green Chemistry: Theory and Practice* (Anastas and Warner 2000). Despite a slow adoption by industry, today this concept is a key pillar in sustainability (see also Anastas and Eghbali 2010).

The concept of green chemistry has become more ingrained in sustainability thinking partly because of use of incentives and the recognition of green chemistry product successes. The Presidential Green Chemistry Award, a joint effort between EPA and the American Chemical Society with additional support provided by other chemical organizations, is an example of using recognition to spur further research and development in this field. This award was first presented in 1996 and has since been given annually to winners who have created innovative chemical technologies that prevent pollution and have broad applicability in industry. Winners are awarded by a jury empanelled by EPA in five categories: small business, academic, and three industry categories with varying focus areas. This award, which has successfully promoted green and sustainable chemistry research and commercialization, is now one of the highest honors available to chemists, chemical engineers, academics, and companies trying to make a significant difference through science and technology in promoting a sustainable world.

to discount new methods as too expensive and risky and a tendency to duplicate infrastructure (e.g., installing a full stormwater system under a swale system).

Finally, it is useful for EPA to examine the extent to which its governing laws provide opportunities or obstacles to achieving sustainability. A prominent example of the removal or lowering of a barrier involves the landowner liability provisions of the 1980 CERCLA (Superfund) statute. Originally established to impose liability on persons who used or allowed their land to be used for improper management of hazardous substances, the landowner liability provision of CERCLA became an obstacle to the sale and redevelopment of these properties. State and federal programs now provide mechanisms to assist both the remediation and the redevelopment of these properties. These programs foster more sustainable development in urban areas, where many of the contaminated sites exist and where there is already infrastructure (e.g., roads, sewer, and water service) and a close-by workforce to support development, rather than in "greenfields," where there is no existing infrastructure and the available workforce may live some distance away (Eisen 2002, 2009). The agency would benefit from identifying any other legal barriers to sustainability and remove or help to remove them to promote sustainability.

FINDINGS AND RECOMMENDATIONS

6.1. Key Finding: Integrating sustainability into the agency's work and culture will be most effective when based on clear principles, vision, strategic goals, and implementation processes (p.43-45).

6.1. Key Recommendation: EPA should institute a focused program of change management to achieve the goal of incorporating sustainability into all agency thinking to optimize the social, environmental, and economic benefits of its decisions, and create a new culture among all EPA employees.

6.2. Finding: Although EPA has developed tools and significant knowledge to implement the Sustainability Framework, further innovations will be necessary to meet the increasingly complex challenges it faces (p.59-65).

6.2. Recommendation: The committee recommends EPA foster innovation at all levels of the organization to meet the challenges of increasingly complex problems.

6.3. Finding: The agency can accelerate its cultural change by practicing sustainability in its day-to-day operations (p.94-95).

6.3.1. Recommendation: Consistent with Executive Order 13514, EPA should implement an internal program to identify key sustainability indicators and associated metrics and implement a tracking and reporting system to demonstrate progress toward the goals of more sustainable operational practices and benchmark performance against other federal or government agencies and private-sector organizations (e.g., nongovernmental organizations [NGOs]).

6.3.2. Recommendation: The agency should at the regional and headquarters level regularly produce a sustainability report on its operations by using widely recognized metrics (such as those of the Global Reporting Initiative).

6.4. Finding: Communicating the basic elements of sustainability to EPA staff is an important step in achieving the cultural change (p.94-95).

6.4. Recommendation: The committee recommends EPA undertake an employee communication and education program to produce the cultural change necessary to support the adoption of the Sustainability Framework and to enable employees at all levels to participate in its implementation.

6.5. Finding: Early pilot efforts in the use of sustainability in a well-defined geographic area will facilitate change in EPA's headquarters (p.97-101).

6.5. Recommendation: EPA should consider a set of place-based and programmatic-based pilots to develop expertise, encourage the cultural change necessary, and demonstrate value for the successful adoption of the Sustainability Framework.

6.6. Finding: The committee supports the refocusing of EPA's ORD as part of the agency's turn toward sustainability. The committee recognizes that further reorganization may be the consequence of the further pursuit of sustainability but that reorganization at the federal level is a process that requires careful consideration (p.96-97).

6.6. Recommendation: The incorporation of a culture of sustainability within the operations of the agency is essential. Managing this cultural change should take precedence over organizational change.

6.7. Finding: To change the direction of an organization, its budgets must reflect that change (p.96).

6.7. Recommendation: The agency's plans and budgets should make room for national and regional work to test for and implement sustainability initiatives.

6.8. Finding: EPA needs to remove barriers to the promotion of sustainability. EPA leadership has long recognized that its organizational structure built around the regulation of specific laws has led to stove-piping and other obstacles to maximizing the benefits of effective and efficient integration of agency actions (p.95).

6.8.1. Recommendation: EPA should examine the extent to which its governing laws provide opportunities or obstacles to the achievement of sustainability.

6.8.2. Recommendation: To accomplish sustainability, greater emphasis in the R&D planning and budgeting process needs to be placed on long-term and multimedia activities.

6.9. Finding: EPA needs to utilize partnerships to incorporate sustainability at all levels (p.92).

6.9.1. Recommendation: The committee recommends EPA maximize opportunities to collaborate with other federal agencies, state and local governments, NGOs, and the private sectors by promoting partnerships that ensure EPA is both a voice at the table and a driver for full and careful consideration of sustainability issues in major national initiatives. EPA should also ensure that an optimal mix of local, regional, and federal agencies are at the table in community and other place- and programmatic-based decision making and ensure that other stakeholders, including NGOs and business groups, are involved.

6.9.2. Recommendation: EPA should, in partnership with appropriate stakeholders, incorporate sustainability in its public education campaigns and encourage other agencies and organizations to adopt a sustainability approach.

6.9.3. Recommendation: EPA should continue to work with other federal agencies to address and implement sustainability research.

6.10. Key Finding: Incorporating sustainability into EPA's mode of operation will require a shift in thinking toward a more systems-based approach that integrates multiple media with multiple objectives in social (including health), environmental, and economic pillars and that considers both short-term and long-term consequences. To accomplish that, EPA will need expertise across many disciplines in implementing a sustainability framework (p.95).

6.10. Key Recommendation: The committee recommends that EPA hire multidisciplinary professionals who are proficient in many disciplines, have experience in the development and implementation in the sustainability assessment tools described, and have a working knowledge in all three pillars and their application to environmental issues. The agency should hire leaders and scientists including from outside sectors to aid the agency in shifting to a more cross-cutting mind-set. Although EPA has existing staff in all the main areas of sustainability-related fields, the agency should further facilitate collaboration among existing professional expertise to encourage dialogue and understanding of the various fields and work already being done within EPA.

REFERENCES

Anastas, P.T., and N. Eghbali. 2010. Green chemistry: Principles and practice. Chem. Soc. Rev. 39(1):301-312.
Anastas, P.T., and J.C. Warner. 2000. Green Chemistry: Theory and Practice. New York: Oxford University Press.

Bruch, C., F. Irwin, and G.D. Bass. 2009. Public access to information, participation, and justice: Forward and a backward steps toward an informed and engaged citizenry. Pp. 459-478 in Agenda for a Sustainable America, J.C. Dernbach, ed. Washington, DC: Environmental Law Institute.

CERP (Comprehensive Everglades Restoration Plan). 2010. CERP 2010 Report to Congress. U.S. Army Corps of Engineers, Jacksonville District, and U.S. Department of the Interior [online]. Available: http://www.evergladesplan.org/pm/program_docs/cerp_reports_congress.aspx [accessed May 2, 2011].

City of Philadelphia. 2011. Retrofit 15 Percent of Housing Stock With Insulation, Air Sealing and Cool Roofs. The Mayor's Office of Sustainability, City of Philadelphia [online]. Available: http://www.phila.gov/green/greenworks/energy_target3.html [accessed June 16, 2011].

DeShazo, J.R., R. Matulka, and N. Wong. 2011. Los Angeles Solar Atlas. UCLA School of Public Affairs, Luskin Center for Innovation [online]. Available: http://luskin.ucla.edu/content/-los-angeles-solar-atlas [accessed June 16, 2011].

DOE (U.S. Department of Energy). 2010. Building Technologies Program: Cool Roofs. Energy Efficiency and Renewable Energy, U.S. Department of Energy [online]. Available: http://www1.eere.energy.gov/buildings/cool_roofs.html [accessed June 16, 2011].

DOE/EPA (U.S. Department of Energy and U.S. Environmental Protection Agency). 2011. Energy Star. U.S. Department of Energy and U.S. Environmental Protection Agency [online]. Available: http://www.energystar.gov/ [accessed June 16, 2011].

Eisen, J.B. 2002. Brownfields redevelopment. Pp. 457-466 in Stumbling Toward Sustainability, J.C. Dernbach, ed. Washington, DC: Environmental Law Institute.

Eisen, J.B. 2009. Brownfields redevelopment: From individual sites to smart growth. Pp. 57-70 in Agenda for a Sustainable America, J.C. Dernbach, ed. Washington, DC: Environmental Law Institute.

Elzen, B., and A. Wieczorek. 2005. Transitions towards sustainability through system innovation. Technol. Forecast. Soc. Change 72(6):651-661.

EPA (U.S. Environmental Protection Agency). 2001. EPA Guidance: Improving Air Quality Through Land Use Activities. EPA420-R-01-001. Office of Air and Radiation, Office of Transportation and Air Quality, U.S. Environmental Protection Agency. January 2001 [online]. Available: http://www.epa.gov/oms/stateresources/policy/transp/landuse/r01001.pdf [accessed May 5, 2011].

EPA. 2007. Sustainability Research Strategy. EPA 600/S-07/001. Office of Research and Development, U.S. Environmental Protection Agency, Washington, DC. October 2007 [online]. Available: http://www.epa.gov/sustainability/pdfs/EPA-12057_SRS_R4-1.pdf [accessed May 3, 2011].

EPA. 2010a. EPA Science Resources. Region 9: Regional Science Council, U.S. Environmental Protection Agency [online]. Available: http://www.epa.gov/region09//science/sci-links.html [accessed July 14, 2011].

EPA. 2010b. Partnership for Sustainable Communities: A Year of Progress for American Communities. EPA 231-K-10-002. Office of Sustainable Communities, U.S. Environmental Protection Agency. October 2010 [online]. Available: http://www.epa.gov/smartgrowth/pdf/partnership_year1.pdf [accessed July 14, 2011].

EPASAB (U.S. Environmental Protection Agency Science Advisory Board). 1988. Future Risk: Research Strategies for the 1990s. SAB-EC-88-040. Science Advisory Board, U.S. Environmental Protection Agency, Washington, DC.

EPASAB. 2010. Science Advisory Board Comments on the President's Requested FY 2011 Research Budget. EPA-SAB-10-005. Science Advisory Board, U.S. Environmental Protection Agency, Washington, DC [online]. Available: http://yosemite.epa.gov/sab/sabproduct.nsf/7867CB8F65EFBC33852576F600418A43/$File/EPA-SAB-10-005-unsigned.pdf [accessed May 3, 2011].

Hecht, A. 2009. Government perspectives on sustainability. CEP 105(1):41-46.

Jackson, L.P. 2010. Letter to Deborah L. Swackhamer, Chairwomen, Science Advisory Board, from Lisa P. Jackson, Administrator, U.S. Environmental Protection Agency, Washington, DC, September 27, 2010.

Kemp, R., D. Loorbach, and J. Rotmans. 2007. Transition management as a model for managing processes of co-evolution towards sustainable development. Int. J. Sustain. Dev. World Ecol. 14(1):78-91.

Lindenmayer, D., R.J. Hobbs, R. Montague-Drake, J. Alexandra, A. Bennett, M. Burgman, P. Cale, A. Calhoun, V. Cramer, P. Cullen, D. Driscoll, L. Fahrig, J. Fischer, J. Franklin, Y. Haila, M. Hunter, P. Gibbons, S. Lake, G. Luck, C. MacGregor, S. McIntyre, R.M. Nally, A. Manning, J. Miller, H. Mooney, R. Noss, H. Possingham, D. Saunders, F. Schmiegelow, M. Scott, D. Simberloff, T. Sisk, G. Tabor, B. Walker, J. Wiens, J. Woinarski, and E. Zavaleta. 2008. A checklist for ecological management of landscapes for conservation. Ecol. Lett. 11(1):78-91.

McKenzie-Mohr, D., and W. Smith. 1999. Fostering Sustainable Behavior: An Introduction to Community Based Social Marketing. Gabriola Island B.C., Canada: New Society Publishers.

NAE (National Academy of Engineering). 1994. The Greening of Industrial Ecosystems, B.R. Allenby, and D.J. Richards, eds. Washington, DC: National Academy Press. 272 pp.

NAPA (National Academy of Public Administration). 1994. A Review, Evaluation, and Critique of a Study of EPA Laboratories by the MITRE Corporation and Additional Commentary on EPA's Science and Technology Programs.

Nill, J., and R. Kemp. 2009. Evolutionary approaches for sustainable innovation policies: From nice to paradigm? Res. Policy 38(4):668-680.

NRC (National Research Council). 1997. Building a Foundation for Sound Environmental Decisions. Washington, DC: National Academy Press.

NRC. 2000a. Strengthening Science at the U.S. Environmental Protection Agency: Research-Management and Peer-Review Practices. Washington, DC: National Academy Press.

NRC. 2000b. Watershed Management for Potable Water Supply: Assessing the New York City Strategy. Washington, DC: National Academy Press.

NRC. 2009. Enhancing the Effectiveness of Sustainability Partnerships: Summary of a Workshop. Washington, DC: The National Academies Press.

NRC. 2010. The Use of Title 42 Authority at the U.S. Environmental Protection Agency: A Letter Report. Washington, DC: The National Academies Press.

NRC. 2011. Progress toward Restoring the Everglades: The Third Biennial Review—2010. Washington, DC: The National Academies Press.

OSTP (U.S. Office of Science and Technology Policy). 1994. Technology for a Sustainable Future: A Framework for Action. U.S. Office of Science and Technology Policy, Washington, DC.

Powell, M.R. 1999. Science at EPA: Information in the Regulatory Process. Washington, DC: Resources for the Future.

State of Florida. 2011. Why Restore the Everglades and Lake Okeechobee? State of Florida [online]. Available: http://www.dep.state.fl.us/evergladesforever/restoration/default.htm [accessed May 31, 2011].

Swackhamer, D.L. 2010. Office of Research and Development Strategic Research Directions and Integrated Transdisciplinary Research. Letter to Lisa P. Jackson, Administrator, U.S. Environmental Protection Agency, Washington, DC, from. Deborah L. Swackhamer, Chair, Science Advisory Board, Office of Administrator, U.S. Environmental Protection Agency, Washington, DC. July 8, 2010 [online]. Available: http://yosemite.epa.gov/sab/sabproduct.nsf/E989ECFC125966428525775B0047BE1A/$File/EPA-SAB-10-010-unsigned.pdf [accessed May 3, 2011].

UMN-WRC (University of Minnesota Water Resources Center). 2011. Minnesota Water Sustainability Framework. University of Minnesota Water Resources Center, St Paul, MN [online]. Available: http://wrc.umn.edu/prod/groups/cfans/@pub/@cfans/@wrc/documents/asset/cfans_asset_292471.pdf [accessed May 3, 2011].

UNEP (United Nations Environment Programme). 2007. Cities and Biodiversity: Engaging the Local Authorities During the Enhanced Phase of the Implementation of the Convention on Biological Diversity. Mayors' Meeting on the Contribution of Cities to Achievement of the 2010 Biodiversity Target, March 26-28, 2007, Curitiba, Brazil [online]. Available: http://www.cbd.int/doc/meetings/city/mayors-01/official/mayors-01-02-en.pdf [accessed May 3, 2011].

7

Benefits of a Sustainability Approach at EPA

The essence of the sustainability framework presented in this report is to enable environmental decision makers to consider the longer-term and inter-generational social, environmental, and economic (the "three pillars") impacts. Sustainability means more "comprehensive, farsighted, critical and integrated approaches on important policies, plans, programs, and projects" (Gibson 2006). The framework capitalizes on an increasingly sophisticated suite of tools for as-sessing and forecasting impacts.

This chapter examines the benefits of adopting the Sustainability Assessment and Management framework in EPA's accomplishment of its mission. It reviews several of the complex challenges the agency faces as it enters its fifth decade of work on behalf of public health and the environment, and as its experience grows in using sustainability principles in decision making. A strong sustainability focus could increase the agency's effectiveness today and over the long term.

A key notion underlying the Committee on Incorporating Sustainability in the U.S. EPA's recommendation for the three pillars and the proposed principles is that EPA needs to move beyond only ensuring that society is "doing less bad" (by applying risk management in regulatory decisions). EPA also needs to ensure that society is "doing more good" at the same time as EPA performs its risk management work and decision-making. Lisa Jackson, EPA administrator, noted that "We have a new opportunity now to focus on how environmentally protective and sustainable we can be. It's the difference between treating disease and pursuing wellness. It's a difference, I believe, that will be fundamental to the future of EPA" (Jackson 2010).

The Sustainability Assessment and Management approach will equip the agency with new tools to analyze and solve new and existing problems and new opportunities to make a positive difference for all Americans, and can also set an example for other federal agencies that face similar challenges to take an integrated approach to public policy issues falling within their purview. These points are amplified in the text below, and Boxes 7-1 to 7-7 provide illustrative examples from the public and private sectors.

DAUNTING CHALLENGES

Despite substantial progress in achieving cleaner air, water, and land in the United States, the nation faces new and even more complex challenges. Environmental quality and natural resources are under great stress from a growing population, ever increasing consumption of energy and natural resources, technologic developments, urbanization, and land development. These pressures are occurring against the background of climate change and its probable disruptive effects on resource productivity, water systems, human health, and ultimately the quality of life and livelihoods on the planet. The scope and complexity of these challenges means that more traditional approaches to environmental protection are not likely to be effective. A selection of complex problems and future risks confronting the agency and the country includes the following:

- Approximately 127 million people lived in counties that exceeded at least one air-quality standard in 2008 (EPA 2010a). Ground-level ozone and particle pollution still exceed health protection levels, and recent scientific studies have established beyond doubt their adverse effects on human heart and respiratory functions (HEI 2010). EPA has a huge number of conventional and toxic pollutant standards to complete as well as its work to begin to control greenhouse gases from stationary sources such as utilities. Multiple pollutant and sectoral strategies and even emissions trading are options under consideration.
- Major nationally important water bodies fail to meet water-quality standards sufficient to protect human uses, such as fisheries and jobs, including the Chesapeake Bay, the Great Lakes, and the Gulf of Mexico. The northern Gulf of Mexico is the site of the second largest dead zone in the world, now measured as the size of New Jersey. The area lacking life-sustaining oxygen (hypoxia) caused by nutrient runoff from the Mississippi River watershed continues to grow. Similar dead zones occur in the Chesapeake Bay and in the coastal waters off the coast of Oregon (Walker 2006, EPA 2010b, NOAA 2010). The U.S. pollution-control system has so far failed to mobilize the resources and actions needed to restore these waters and the marine life and human livelihoods they support.
- Many areas in urban centers, particularly in highly industrialized zones, contain persistent sources of contamination due to past disposal prac-

tices. Although cleanup has occurred, having over 450,000 brownfield sites (see EPA 2010c), many sites remain as potential sources of human and environmental exposure. Cleanup of large complex sites remains difficult and the cleanup process is slow. Only 347 Superfund sites of the roughly 2,000 sites listed in the National Priorities List (NPL) have been taken off the list (as of March 2011), and the 347 sites still contain residual contamination that may pose long term risks to the environment.

- The cumulative burden of pollution sources, even those meeting standards, on communities already lacking adequate housing, health care, and other community services requires attention to environmental equity and a search for innovative ways to provide benefits to community residents in site remediation and redevelopment decision making. EPA and its regions have been pioneers in making environmental justice a priority in their work with communities and other federal agencies to deliver multiple benefits.

- Climate change presents significant risks to the environment, human health, and society. "Climate change is occurring, is very likely caused largely by human activities, and poses significant risks for—and in many cases is already affecting—a broad range of human and natural systems" (NRC 2010a, p.3). Projected impacts on the United States include greater risk of drought in the West and increased flooding and reduced water quality in most regions; adverse effects on crops and livestock production; increased risk in coastal areas of sea-level rise and storm surge; and greater heat-stress and other human health risks (USGCRP 2009). Because climate change is affecting—or will affect—every sector of society and because climate-change mitigation and adaptation will require all levels of government and society as well as a great variety of legal and policy tools, traditional pollution-control approaches by themselves will be insufficient (NRC 2010a, b, c, d).

- Emerging technologies are hard for agencies like EPA to evaluate and oversee using existing legal mandates and analytic tools. Nanomaterials, materials so small that they require super-powerful microscopes to be seen, are likely to become a major part of the economy in the coming decades. Already there are more than 1,000 consumer products that use such materials and probably an even larger number of industrial products that use them. Standard risk-based approaches to hazard identification and dose-response analyses are challenged by agents that have different properties at such a small size and agents that may be more toxic at lower doses (Goldstein 2010). The whole area of nanotechnology is one which requires the anticipatory approach encouraged by sustainability. Sound decisions about nanomaterials will require consideration of social and economic factors as well as environmental ones (Davies 2006, 2009; Hodge et al. 2011).

PROBABLE BENEFITS OF A MORE ROBUST
APPROACH TO SUSTAINABILITY

Given this brief inventory of some of today's pressing problems, there is wide agreement that the United States needs to find new solutions to achieve objectives set forth in EPA's governing statutes and to extend the benefits of a clean environment to everyone. EPA is already active in applying sustainability thinking to environmental problem solving in particular cases. Boxes 7-1, 7-2, and 7-7 in this chapter describe the kinds of activities or problem solving EPA has already undertaken. They provide evidence of the benefits that can be gained from a broader analysis of the social, environmental, and economic impacts of alternatives and from a collaboration with other agencies and stakeholders on finding solutions. The benefits likely to result from greater incorporation of sustainability into EPA's work include the following:

- *Reduced compliance costs.* The use of green infrastructure in cities with combined sanitary and storm-sewer overflows in Kansas City, Philadelphia, and elsewhere has already saved cities hundreds of millions of dollars. Similarly, installation of energy-saving light fixtures in more than 700 schools in the New York City public-school system made it possible to replace existing and contaminated fixtures within the system's existing budget rather than requiring it to spend more than 1 billion dollars. Energy efficiency and conservation can also reduce greenhouse gas emissions and produce cost savings at the same time. By combining otherwise different regulatory regimes, EPA, DOT, and the state of California reduced compliance costs for greenhouse gas and fuel economy rules for automobile manufacturers.
- *Greater environmental justice and more livable communities.* The EPA-HUD-DOT collaboration in Boston's Fairmont Rail Corridor will clean up contaminated sites, create jobs, provide new housing, and improve public access to mass transit. The use of green infrastructure tends to make communities more attractive. The reconstruction of a public bus depot in northern Manhattan to a "green" bus depot helped to reduce air pollution.
- *Greater environmental and public health benefits.* Philadelphia's use of green infrastructure is also protecting its drinking-water sources. The use of energy-efficient lighting in New York City public schools is making it possible for the city to replace its existing PCB contaminated fluorescent lights, reducing student and teacher exposure to a known carcinogen. The work of the SURF, which EPA supports, will probably lead to more sustainable remedial decisions while meeting the protective cleanup standards at many sites.
- *More effective use of federal funds.* By pooling resources to address sustainability problems in the Fairmont Rail Corridor, three federal

BOX 7-1
Green Infrastructure: Sustainable Water Quality Solutions for Cities with Combined Sewer and Storm-Sewer Overflows

In many older cities in the northeast and Midwest, the most serious water-quality problems result from the overflow of sewers containing untreated sewage when rainwater enters combined sewers and the resulting flow exceeds the capacity of the wastewater treatment plant. The construction of separate pipes or tunnels and storage capacity to hold water until it can be directed to the treatment plant involves costly engineering and construction projects and substantial disruption to streets and businesses for long periods. These costs and community impacts have led to more serious consideration of green infrastructure approaches that could reduce the inflow of rain and stormwater runoff through nonstructural solutions.

EPA defines green infrastructure projects as wastewater treatment technologies or processes that use natural or engineered systems, such as green roofs, rain gardens, and permeable placement, that mimic natural processes and direct stormwater to areas where it can be stored, infiltrated, evapotranspirated, or reused. There are encouraging examples of more willingness to test green and mixed gray and green solutions to achieve more cost-effective and environmentally beneficial results. For example, the city of Portland, Oregon built tunnel and storage facilities as required by the state permitting authority but also installed over 400,000 square feet of green roofs and is adding over 80,000 trees, all to manage stormwater as close to the source as possible and to test the approach for future growth. Last year, in Kansas City, Missouri, EPA settled a major case of water-pollution violations, which provides a mix of control approaches, including an adaptive management approach, to use green infrastructure effectively in lieu of or in addition to structural controls (source: EPA consent decree with Kansas City [2010]).

These examples help demonstrate how green systems will actually work and how the adoption of experimental approaches can satisfy the requirements of EPA's combined sewer overflow standards. They offer the promise of less costly and more sustainable urban water solutions achieved through more flexible and collaborative approach (Scarlett 2010).

agencies are likely to achieve greater benefits than if they had addressed these problems separately or in separate places.

Companies that successfully adopted sustainability strategies can offer some useful approaches and tools to EPA (Boxes 7-5 and 7-6). Corporations are in the business of making goods and services for customers for a profit; providing additional social and environmental benefits is not normally considered to be their primary mission. Beginning with the Rio Earth Summit in 1992, however, forward-looking businesses, led by the World Business Council for Sustainable Development, have worked to define and implement principles of sustainability in the belief that they will make their companies more competitive, resilient,

BOX 7-2
Using EPA Technical Assistance to Aid Advances in
Stormwater Best Practices

In cooperation with EPA, the city of Philadelphia has implemented select green infrastructure features to manage stormwater runoff more efficiently. The Philadelphia Water Department has implemented green infrastructure through demonstration and restoration projects, a new stormwater fee system, and new stormwater regulations for all new construction and redevelopment projects. Philadelphia's stormwater fee structure is now based on amount of square feet of a property's impervious surface, allowing the financial burden of the fees to fall more equitably on parties that own the greatest amount of contributing surface. Philadelphia's permitting system was restructured to streamline stormwater permits, easement of flood and channel requirements if 20% reduction in impervious surface can be met. As a result, most developers now build on infill sites instead of undeveloped natural areas. Many of these urban redevelopment projects reach the 20% reduction through a variety of green infrastructure practices, including roof downspout disconnections, porous pavement, tree plantings, and green roofs. These areas manage most 1-inch rain events, reduce CSO inputs by a quarter billion gallons, and have saved Philadelphia an estimated $170,000,000 in infrastructure costs. The city also saves money on the costs of maintaining pipe networks and the upkeep of treatment plants by removing flow from these systems. Through interagency and stakeholder collaboration, fiscal responsibility, and a commitment to sustainable development best practices, Philadelphia has benefited from green infrastructure by improved compliance with the National Pollutant Discharge Elimination System and the Total Maximum Daily Load requirements of the Clean Water Act, increased protection of drinking-water sources, and aesthetic improvements in the urban built environment (EPA 2009).

nimble, able to attract and retain customers and employees, and better able to work with government regulators and financial institutions. The World Business Council for Sustainable Development, where much of this work began, recently produced "Vision 2050," which outlined strategies to take the economy and the market to a far more sustainable future (WBCSD 2010).

In addition to considering the experience and lessons learned by companies in their sustainability programs, EPA has the opportunity as a science agency to support company efforts by working collaboratively with industry on issues of mutual interest, such as defining best practices in life cycle analysis, where there are many different tools and applications available for users. EPA's focus and expertise in green chemistry is another sustainability interest that is shared by industry and presents opportunities for further exploration of ways to make green chemistry principles more widely understood and adopted. Hecht (2009) noted that "EPA has been active domestically and internationally in promoting sustainable development through green chemistry." The committee is confident that a

BOX 7-3
Energy Savings Permit Cleanup of Polychlorinated Biphenyl
(PCB) Contamination in New York City Schools
Without Layoff of Teachers

In response to concerns about PCB contamination in light fixtures in public schools, New York City has agreed to a broad cost-saving effort to replacing these faulty fixtures with energy-efficient alternatives (NYCDOE 2011). EPA investigation of older fluorescent light fixtures has shown leaking capacitors and frequent exceedences of PCB guidelines in virtually every New York City school thus far evaluated (EPA 2011). PCBs are considered to be human carcinogens, and leaking capacitors have resulted in electrical fires. The New York City public-school system initially resisted the call to replace the light fixtures present in over 700 schools because it would cost over a billion dollars and lead to the layoff of teachers. However, when the energy saving of the new light fixtures was included in the estimates, New York City found that the replacement costs were low enough to be accommodated within its usual capital budget without any impact on school operations (Navarro 2011). In addition to improving efficiency and cost-savings in the long run, school buildings will undergo regular energy audits to determine how to improve overall energy efficiency in each building. Improvements will translate into even greater cost savings in the long run (NYCDOE 2011).

well-defined sustainability framework to support priority setting and decision making at the agency can, over time, multiply the cases of optimizing solutions to problems and take EPA's performance to a new level. As illustrated with the text boxes in this report, such a framework has the potential not only to mitigate harmful unintended consequences of decisions but also to foster more imagination and creativity in forging solutions to pressing problems.

CONCLUSION: THE JOURNEY AHEAD

A sustainability approach can strengthen EPA as an organization and a leader in the nation's progress toward a sustainable future. Adopting a vision for sustainability as a goal will provide a unifying and forward-looking stimulus to the agency. EPA has never had a formal mission in law. Its individual legal mandates, which have driven the agency's actions and analysis, have not been revised in many years. The individual programs have lacked a vision as a basis for a unified approach. Sustainability gives the agency a more positive vision and lays the basis for an approach capable of inspiring commitment and enlisting cooperation from different parts of the agency as well as from diverse stakeholders.

As the vision of sustainability is implemented, the processes that follow from the vision bring a variety of benefits. EPA and its programs can be organized on the basis of approaches that cut across the traditional boundaries enshrined in law.

BOX 7-4
Sustainable Solutions to Air Pollution Associated with
Public Transit Bus Depots in Northern Manhattan:
An Environmental Justice Issue and Community Response

WE ACT, an environmental justice organization, has demonstrated the success of community-based organizations in encouraging sustainable efforts to address environmental health issues in disadvantaged communities. Air pollution associated with idling buses at transit depots in Manhattan has been an important concern for neighboring communities, particularly as studies have indicated that traffic exposure exacerbates asthma among children (HEI 2010). Five of six public transit bus depots in Manhattan are located in low-income communities. In 2000, following a complaint by WE ACT, the Federal Transit Administration ruled that the Metropolitan Transit Authority (MTA) was discriminatory in their siting of the depots and had failed to comply with rules to promote public health and community development. Ultimately, the MTA decided to remove and rebuild one of the facilities, the Mother Clara Hale Bus depot. WE ACT partnered with MTA to support sustainability efforts at this redevelopment, including work toward a Leadership in Energy and Environmental Design (LEED) green building certification. In addition, WE ACT worked to increase community involvement in the project by soliciting advice about the redevelopment of the depot and organizing the Mother Clara Hale Community Taskforce to formalize the community's role in development activities. In 2008, the Mother Clara Hale Community Taskforce recruited 170 residents to assist MTA in designing the depot with green design features, such as a green roof, gray-water reclamation, and air-pollution controls The net result is an involved community now cooperating with local government in decreasing air pollution while increasing sustainability (WE ACT 2010).

Place-based efforts can increasingly become the ways in which agency programs are implemented.

Central to the integrated and positive sustainability approach is a focus on preventing harm rather than dealing with it after it occurs. Pollution prevention under the Sustainability Framework would not be categorized as belonging to any single medium, and would aim to be socially beneficial, environmentally effective, and economically more efficient than current pollution control. Sustainability encourages EPA to anticipate problems and invest in solutions before the problems become critical. Encouraging a more long-range and holistic view, sustainability is likely to enable EPA to avoid mistakes and thus maximize the value received for the agency's investments.

EPA's efforts can become more internally coordinated and its sustainability approach can become more inclusive of other regulatory and nonregulatory agencies, state and local agencies, businesses, and nongovernmental organizations. It will not be possible for EPA to incorporate social and economic factors without

---

BOX 7-5
The Sustainability Remediation Forum: A Private Sector Effort
to Incorporate Sustainable Practices into Remediation Efforts

   Large remediation projects are energy and resource intensive and may ultimately result in discharges of various air pollutants, including greenhouse gases. The Sustainable Remediation Forum (SURF), initiated in 2006 by DuPont, provides an example of a private sector effort to incorporate sustainability into decisions on remedial actions at contaminated sites. The mission of this forum, which has received some support from EPA (Region 2), is to "establish a framework that incorporates sustainable concepts throughout the remedial action process while continuing to provide long-term protection of human health and the environment and achieving public and regulatory acceptance" (Ellis and Hadley 2009). SURF, which has grown in size to over 300 members including the international community, recently published a white paper providing recommendations on how to integrate sustainability principles into remediation projects. SURF noted that sustainability can be applied both to new projects and those already under way, adding that the remediation profession needs to consider sustainability principles and practices in all remediation-related activities. The group also noted that that sustainability assessments matter during remediation, recommending that these assessments be flexible, site-specific, and reflect stakeholder values. Uniform guidelines and metrics are necessary to produce consistent sustainability assessments (Ellis and Hadley 2009). In addition to SURF, the Interstate Technology and Regulatory Commission (ITRC) has recently published a document titled "Green and Sustainable Remediation: State of the Science and Practice (ITRC 2011). In addition to a comprehensive evaluation of technologies for more sustainable soil and groundwater cleanups, the document provides an appendix summarizing over 30 software tools that can be used to conduct sustainability assessment of remedial options.

---

drawing on many non-EPA actors. At the same time, sustainability will impose a leadership role on EPA because the agency may need to take the lead in convincing other agencies and organizations to incorporate environmental considerations in nonenvironmental decisions. The agency may have to devote resources to providing technical assistance to make such incorporation effective. It also may have to set an example in adopting sustainability by providing leadership at the national level.

   The sustainability mandate to consider social and equity factors can make EPA even more sensitive than it has been to populations that are disproportionately exposed to environmental risks. The agency has a long history and has developed procedures for incorporating environmental justice in its decisions (EPA 2010d,e). Sustainability can reinforce these procedures and will provide a logic and framework for considering environmental justice as part of every major decision.

BOX 7-6
IBM: Early Mover on Corporate Responsibility and Sustainability

IBM, a company with revenues approaching $100 billion, believes that focusing on the idea of a smarter planet is beneficial to its business growth and profitability as well as advancing its values. IBM's environmental policy, first established in 1971, is part of the company's long-standing, broad-based objective to be at the forefront of companies working to make the world a better place. Company Chairman and CEO Samuel J. Palmisano summed up the IBM position in its 2009 report: "Addressing the issues facing the world now—from clean water, better health care, green energy, and better schools to sustainable and vibrant cities and an empowered workforce and citizenry—does not pose a choice between business strategy and citizenship strategy. Rather, it represents a fusion of the two" (IBM 2010, p. 1). The company's environmental sustainability priorities, developed by analyzing external stakeholder interests are as follows:

- Energy conservation and climate protection
- Process stewardship
- Product stewardship
- Supply chain management

These objectives often involve actions not required by U.S. law. IBM's annual energy conservation goal to conserve energy in an amount equivalent to 3.5% of IBM's energy use was exceeded, reaching 5.4%. Cuts in carbon dioxide (CO_2) emissions have been a priority since the early nineties. From 1990 through 2009, IBM saved over 5.1 billion kilowatt hours of energy consumption, avoided over 3 million tons of CO_2 emissions, and saved over $370 million. IBM has since embarked on a second generation energy objective under which its CO_2 emissions from a 2005 baseline have decreased by a further 5.7%. In the area of process stewardship, by January 31, 2010, IBM had eliminated all uses of perflourooctane sulfonate and perflourooctanoic acid, two persistent chemicals from its microprocessing manufacturing processes.

The company has also increased the use of recycled plastics in its products and reduced packaging material. Its researchers are working with Stanford University scientists to develop green chemistries that could result in biodegradable products made from renewable resources. Success will mean that plastic bottles recycled once but then disposed of in landfills might degrade.

Supply-chain management strategies are focused on guiding the capacity and accountability of IBM's suppliers to succeed. In 2010, IBM announced a first-of-a-kind requirement for its global suppliers to develop environmental management systems, establish their own goals, and publicly disclose their progress. Identifying use of "conflict materials" (e.g., timber or diamonds used to fund civil wars) and their sources is another new initiative responsive to both social and environmental concerns (IBM 2010).

BOX 7-7
Climate Change Mitigation and Sustainability

For more than a decade, state laws to develop renewable energy and energy efficiency have been known to create jobs, result in the development of new technologies, reduce the effect of high and fluctuating fossil-fuel prices on the poor as well as business, lead to economic savings, and reduce pollutants, such as sulfur dioxide and particulates. The laws have also, and often incidentally, reduced greenhouse gas emissions (Dernbach et al. 2000). As states have begun to address climate change more directly, it is apparent that appropriate legal and policy instruments, if scaled nationally, could create millions of net new jobs, produce considerable economic savings, create a net increase in gross domestic product, and significantly reduce U.S. greenhouse gas emissions (Peterson and Wennberg, 2010). Perhaps the most important of these instruments are those that foster energy conservation and efficiency (NRC 2010e).

At the federal level, EPA and the U.S. Department of Transportation (DOT) in 2010 adopted combined corporate average fuel efficiency (CAFE) standards and greenhouse gas emission limits for light-duty motor vehicles (including passenger cars and light-duty trucks). Beginning in 2016, these vehicles will be required to have a combined average emissions level not exceeding 250 grams of CO_2 per mile, which is equivalent to 35.5 miles per gallon. The government estimates that "these standards will cut greenhouse gas emissions by an estimated 960 million metric tons and 1.8 billion barrels of oil over the lifetime of the vehicles sold under the program." This rule making brought together a great variety of stakeholders, including automobile manufacturers and the state of California. In prior regulatory efforts, EPA, California, and DOT adopted separate standards for automobile pollution and fuel efficiency. By contrast, this rule represents a harmonized national program for a new generation of cleaner vehicles that is explicitly intended to foster economic development, job creation, environmental protection, and national security (EPA/DOT 2010).

The emphasis on the future contained in sustainability's concern for future generations is likely to have an impact on EPA's perspective as well as on its decision-making process. The agency can become more anticipatory, making greater use of new science and of forecasting. At the same time, it may become more self-evaluating, making greater use of health and environmental monitoring tools, and program evaluation tools, such as benefit-cost analysis and life-cycle analysis (Chapter 4).

There is no certainty that these changes will take place in the agency or to what degree they will take hold or what form they will take. But the stimulus provided by adopting a sustainability framework could provoke reflection and change within EPA, and the changes will better equip the agency to deal with the challenges it will face. Assuming that EPA adopts the goal of sustainability, there will be benefits for the United States as a whole. There is likely to be a closer meshing of economic and environmental policies to the benefit of both. The result is likely to be a cleaner environment and stronger economy. These benefits are likely to be shared more equitably as social considerations are weighed alongside environmental and economic ones. The economic benefits may not be limited to improved coordination. As sustainability focuses attention on greener products, the United States can hope to capture a larger part of the world market in a variety of goods. It can play an important role in building the tools and workforce for a more sustainable world.

In summary, the committee sees EPA moving into a leadership role in using a sustainability framework to deliver better results for the nation and its future. EPA's national pollution-control responsibilities give the agency important influence on management of natural resources and the opportunity and obligation to help build better national policy and strategy. As an agency with strong science capability and wide expertise, EPA is equipped to be a catalyst for sustainability in activities beyond its traditional regulatory and grant-making roles. EPA is a thought leader, convener of governmental partners and private stakeholders, and funder of innovative environmental strategies; and the agency provides reliable information on environmental conditions and materials that can inform government, businesses, citizens, and consumers in their choices and decisions. These capabilities provide a solid platform for EPA leadership in advancing a sustainability agenda.

FINDINGS AND RECOMMENDATIONS

7.1. Finding: EPA can adopt sustainability principles while fulfilling its core mission of protecting public health and the environment. The embrace of sustainability at EPA can promote new thinking and provide powerful new tools to forge better solutions to current and future problems (p.112-113).

7.1. Recommendation: EPA should foster a culture of sustainability to increase EPA's capacity to imagine and implement better solutions and increase recognition of the economic and social value of the benefits of environmental protection. Agency staff should be encouraged to seek opportunities to further EPA's sustainability goals in all decisions and actions.

7.2. Finding: The agency can be a promoter of economic development as well as a regulator. In that role, for example, EPA can help to encourage manufacturing of greener products and to create new markets in sustainable goods and services (p.114-117).

7.2. Recommendation: EPA expertise in sustainability should be leveraged into nonregulatory environmental improvement programs for businesses of all sizes, creating synergy for the sustainability, public health, and competitiveness of American businesses.

7.3. Finding: Disadvantaged communities often bear a disproportionate burden of environmental stressors, such as higher pollutant burdens that are often coupled with poor housing, inadequate health care, and other place-based problems (p.114).

7.3. Recommendation: To maximize social benefits as well as reduce health risks, EPA should target activities to decrease and eliminate environmental inequities. Research aimed at elucidating the cause-and-effect relationship between an environmental problem and an adverse consequence, especially cumulative impacts, should be focused on disadvantaged communities and should seek their engagement and cooperation.

REFERENCES

Davies, J.C. 2006. Managing the Effects of Nanotechnology. Washington, DC: Woodrow Wilson International Center for Scholars.

Davies, J.C. 2009. Oversight of Next Generation Nanotechnology. PEN 18. Washington, DC: Woodrow Wilson International Center for Scholars.

Dernbach, J.C. 2000. Moving the climate change debate from models to proposed legislation: Lessons from State experience. Environ. Law Rep. 30(1):10933-10980.

Ellis, D.E., and P.W. Hadley. 2009. Sustainable remediation white paper—Integrating sustainable principles, practices, and metrics into remediation projects. Remed. J. 19(3):5-114.

EPA (U.S. Environmental Protection Agency). 2009. Green Infrastructure: Pennsylvania: Philadelphia. Managing Wet Weather with Green Infrastructure. National Pollutant Discharge Elimination System, U.S. Environmental Protection Agency [online]. Available: http://cfpub.epa.gov/npdes/greeninfrastructure/gicasestudies_specific.cfm?case_id=62 [accessed June 13, 2011].

EPA. 2010a. Our Nation's Air: Status and Trends through 2008. EPA-454/R-09-002. Office of Air Quality Planning and Standards, U.S. Environmental Protection Agency, Research Triangle Park, NC. February 2010 [online]. Available: http://www.epa.gov/airtrends/2010/ [accessed May 3, 2011].

EPA. 2010b. Request for Recommendations on Plan EJ 2014 and Permitting Charge. Memorandum to National Environmental Justice Advisory Council Members, from Cynthia Giles, Assistant Administrator, Office of Enforcement and Compliance Assurance, and Lisa F. Garcia, Senior Advisor to the Administrator for Environmental Justice, Office of the Administrator, U.S. Environmental Protection Agency, Washington, DC.[online]. Available: http://www.epa.gov/compliance/ej/resources/policy/plan-ej-2014.pdf [accessed May 3, 2011].

EPA. 2010c. About Brownfields. U.S. Environmental Protection Agency [online]. Available: http://www.epa.gov/brownfields/about.htm [accessed May 5, 2011].

EPA. 2010d. Interim Guidance on Considering Environmental Justice during Development of an Action. U.S. Environmental Protection Agency, Washington, DC. July 10, 2010 [online]. Available: http://www.epa.gov/environmentaljustice/resources/policy/considering-ej-in-rulemaking-guide-07-2010.pdf [accessed May 3, 2011].

EPA. 2010e. Research Study 'Dead Zones' U.S. Waters to Alleviate Harmful Effects. Office of Research and Development, U.S. Environmental Protection Agency. September 2010 [online]. Available: http://www.epa.gov/ord/npd/pdfs/waterqualityresearch-hypoxia-factsheet.pdf [accessed May 3, 2011].

EPA. 2011. PCBs in Schools. Region 2, U.S. Environmental Protection Agency [online]. Available: http://www.epa.gov/region2/pcbs/index.html [accessed June 16, 2011].

EPA/DOT (U.S. Environmental Protection Agency and U.S. Department of Transportation). 2010. Light-Duty Vehicle Greenhouse Gas Emission Standards and Corporate Average Fuel Economy Standards. EPA-420-R-10-012a. U.S. Environmental Protection Agency and U.S. Department of Transportation, Washington, DC [online]. Available: http://www.epa.gov/oms/climate/regulations/420r10012a.pdf [accessed May 3, 2011].

Gibson, R. 2006. Sustainability assessment: Basic components of a practical approach. IAPA 24(3):170-182.

Goldstein, B.D. 2010. The Scientific Basis for the Regulation of Nanoparticles: Challenging Paracelsus and Pare. UCLA J. Environ. L. Policy 28(1).

Hecht, A. 2009. Government perspectives on sustainability. CEP 105(1):41-46.

HEI (Health Effects Institute). 2010. Traffic-Related Air Pollution: A Critical Review of the Literature on Emissions, Exposure, and Health Effects. Boston, MA: Health Effects Institute [online]. Available: http://pubs.healtheffects.org/getfile.php?u=553 [accessed July 15, 2011].

Hodge, G.A., D.M. Bowman, and A.D. Maynard, eds. 2011. International Handbook on Regulating Nanotechnologies. Cheltenham, UK: Edward Elgar Publishing.

IBM (International Business Machines Corporation). 2010. IBM Corporate Responsibility Report 2009. IBM. June 2010 [online]. Available: http://www.ibm.com/ibm/responsibility/IBM_CorpResp_2009.pdf [accessed May 3, 2011].

ITRC (Interstate Technology & Regulatory Council). 2011. Green and Sustainable Remediation: State of the Science and Practice. GSR-1. Washington, DC: Interstate Technology & Regulatory Council, Green and Sustainable Remediation Team. [online]. Available: www.itrcweb.org [accessed June 20, 2011].

Jackson, L.P. 2010. Remarks to the National Academy of Sciences, November 30, 2010, Washington, DC [online]. Available: http://yosemite.epa.gov/opa/admpress.nsf/8d49f7ad4bbcf4ef852573590040b7f6/1c893e457b3cbb25852577ec0054048c!OpenDocument [accessed Apr. 15, 2011].

Navarro, M. 2011. City to Replace School Lighting Tainted by PCBs. The New York Times. February 23, 2011. [online]. Available: http://www.nytimes.com/2011/02/24/science/earth/24pcb.html?_r=4&ref=mireyanavarro [accessed June 13, 2011].

NOAA (National Oceanic and Atmospheric Administration). 2010. NOAA-Supported Scientists Find Changes to Gulf of Mexico Dead Zone. National Oceanic and Atmospheric Administration, August 9, 2010 [online]. Available: http://www.noaanews.noaa.gov/stories2010/20100809_deadzone.html [accessed May 3, 2011]..

NRC (National Research Council). 2010a. Advancing the Science of Climate Change. Washington, DC: The National Academies Press.

NRC. 2010b. Adapting to the Impacts of Climate Change. Washington, DC: The National Academies Press.

NRC. 2010c. Informing an Effective Response to Climate Change. Washington, DC: The National Academies Press.

NRC. 2010d. Limiting the Magnitude of Future Climate Change. Washington, DC: The National Academies Press.

NRC. 2010e. Real Prospects for Energy Efficiency in the United States. Washington, DC The National Academies Press.

NYCDOE (New York City Department of Education). 2011. City Announces Comprehensive Plan to Increase Energy Efficiency and Environmental Quality at Schools. New York City Department of Education News, February 23, 2011 [online]. Available: http://schools.nyc.gov/Offices/mediarelations/NewsandSpeeches/2010-2011/energyeffandenviroqualityrelease22311.htm [accessed June 16, 2011].

Peterson, T., and J. Wennberg. 2010. Impacts of Comprehensive Climate and Energy Policy Options on the U.S. Economy. Washington, DC: Johns Hopkins University.

Scarlett, L. 2010. Green, Clean and Dollar Smart: Ecosystem Restoration in Cities and Countryside. Washington, DC: Environmental Defense Fund.

USGCRP (U.S. Global Change Research Program). 2009. Global Climate Change Impacts in the
    United States: A State of Knowledge Report from the U.S. Global Change Research Program,
    T.R. Karl, J.M. Melillo, and T.C. Peterson, eds. New York: Cambridge University Press [on-
    line]. Available: http://downloads.globalchange.gov/usimpacts/pdfs/climate-impacts-report.pdf
    [accessed May 3, 2011].
Walker, C. 2006. 'Dead Zone' Off Oregon Coast Is Growing, Study Says. National Geographic News,
    August 4, 2006 [online]. Available: http://news.nationalgeographic.com/news/2006/08/060804-
    dead-zone.html [accessed May 3, 2011].
WBCSD (World Business Council for Sustainable Development). 2010. Vision 2050. World Business
    Council for Sustainable Development, Washington, DC [online]. Available: http://www.wbcsd.
    org/web/projects/BZrole/Vision2050-FullReport_Final.pdf [accessed May 3, 2011].
WE ACT. 2010. Environmental Justice and the Green Economy: A Vision Statement and Case Studies
    for Just and Sustainable Solutions. Roxbury, MA: Alternatives for Community & Environment,
    Inc. [online]. Available: http://www.weact.org/Portals/7/Publications/EJGE_Report_English.
    pdf [accessed May 31, 2011].

Appendix A

The Committee on Incorporating Sustainability in the U.S. Environmental Protection Agency

Bernard Goldstein, M.D., (*Chair*), is professor of environmental and occupational health at the University of Pittsburgh Graduate School of Public Health; he also served as the dean of the Graduate School of Public Health. He was the founding director of the Environmental and Occupational Health Sciences Institute, a joint program of Rutgers, the State University of New Jersey and the University of Medicine and Dentistry of New Jersey (UMDNJ)–Robert Wood Johnson Medical School from 1986 to 2001. He was the chair of the Department of Environmental and Community Medicine, UMDNJ–Robert Wood Johnson Medical School from 1980 to 2001. Dr. Goldstein served as acting dean of the UMDNJ–School of Public Health from 1998 to 1999, the first year of its formation. He is a physician certified by the American Board of Medical Specialties in internal medicine and hematology and in toxicology. He is the author of over 200 articles and book chapters related to environmental health sciences and to public policy. Dr. Goldstein was assistant administrator for research and development, U.S. Environmental Protection Agency (EPA), 1983-1985. His past activities include member and chairman of the National Institutes of Health Toxicology Study Section; the EPA Clean Air Scientific Advisory Committee; and the National Board of Public Health Examiners. He is a member of the Institute of Medicine where he has cochaired the Section on Public Health, Biostatistics, and Epidemiology. He has served as chair or member of numerous Institute of Medicine or National Research Council committees. Dr. Goldstein has also served as president of the Society for Risk Analysis, vice president and editor in chief of the Scientific Committee on Problems of the Environment, and as a member of the National Advisory Environmental Health Sciences Council.

Leslie Carothers is president of the Environmental Law Institute (ELI). ELI is an independent, nonpartisan education and research organization working to protect the environment by improving law, policy, and management. Ms. Carothers has been a professional environmentalist for over 30 years. Before her election as ELI president in June 2003, she served for 11 years as vice president, Environment, Health and Safety at United Technologies Corporation (UTC) in Hartford, a diversified manufacturer of products for the aerospace and building systems markets. Ms. Carothers also served as commissioner of the Connecticut Department of Environmental Protection from 1987 to 1991 and senior environmental counsel for PPG Industries, a manufacturing company in Pittsburgh, from 1982 to 1987. She began her environmental career with EPA in the air pollution program in Washington, DC, in 1971 and later served as enforcement director, deputy regional administrator, and acting regional administrator of the EPA New England Region in Boston. In 1991, she was an adjunct lecturer on environmental regulation at the Yale School of Forestry and Environmental Studies. Ms. Carothers is a past member and chair of the Board of Directors of the Connecticut Audubon Society and ELI and a past member of the Board of the Nature Conservancy (Connecticut chapter). She currently serves on the Board of Directors of Strategies for the Global Environment (Pew Center on Global Climate Change). She is a graduate of Smith College and Harvard Law School and also holds a master's degree in environmental law from George Washington University.

Clarence (Terry) Davies, Ph.D., is a senior fellow at the Resources for the Future. He is a political scientist who, during the last 30 years, has written several books and numerous articles about environmental policy. He chaired the National Research Council (NRC) Committee on Decision Making for Regulating Chemicals in the Environment and was a committee member of the NRC report *Risk Assessment in the Federal Government.* While serving as a consultant to the President's Advisory Council on Executive Organization, he was a coauthor of the reorganization plan that created the EPA. His previous positions have included assistant professor of public policy at Princeton University, executive vice president of the Conservation Foundation, executive director of the National Commission on the Environment, and assistant administrator for policy at EPA. In 2000 he was elected a fellow of the American Association for the Advancement of Science. He received a Ph.D. in American Government from Columbia University and a B.A., cum laude from Dartmouth College.

John Dernbach is distinguished professor of law and director of Environmental Law Center at Widener University School of Law. His scholarship focuses on sustainable development and climate change, and he teaches a variety of courses, including environmental law, international environmental law, sustainability and the law, and climate change. Mr. Dernbach has served as director of the Policy Office at the Pennsylvania Department of Environmental Protection (DEP), which is responsible for developing and coordinating policy and regulatory initiatives for DEP, including the integration of sustainable-development concepts into DEP programs. Over more than a decade at DEP's predecessor agency, the Department of Environmental Resources, he counseled and worked in DEP's mining and waste programs and drafted four laws. Mr. Dernbach has written more than 30 articles for law reviews and peer-reviewed journals and has been an author, coauthor, or contributor of chapters in 13 books. He is the editor of *Agenda for a Sustainable America* (Environmental Law Institute Press, January 2009) and *Stumbling Toward Sustainability* (Environmental Law Institute Press 2002), which are comprehensive assessments of U.S. sustainable-development activities that include recommendations for future efforts. He is a member of the International Union for Conservation of Nature Commission on Environmental Law and served from 2005 to 2008 on the Roundtable on Science and Technology for Sustainability of the National Academy of Sciences.

Paul Gilman, Ph.D., joined Covanta in 2008 as Covanta Energy's first senior vice president and chief sustainability officer. He is responsible for Covanta's safety, health, and environmental compliance programs, and for sustainability initiatives that further reduced Covanta's environmental impact while increasing the use of its technologies. Before joining Covanta, Dr. Gilman was the director of the Oak Ridge Center for Advanced Studies. He served as the assistant administrator for research and development and science advisor at EPA from 2002 until 2004. Prior to joining EPA, he was director for policy planning at Celera Genomics. Dr. Gilman was previously the executive director of life sciences and agriculture divisions of the NRC. In addition, Dr. Gilman has held several senior government positions, including associate director of the White House Office of Management and Budget (OMB) for natural resources, energy, and science, and executive assistant to the secretary of the U.S. Department of Energy for technical matters. He has 13 years of experience working on the staff of the U.S. Senate in several capacities, including as a congressional science fellow sponsored by the American Association for the Advancement of Science.

Neil Hawkins, Ph.D., currently serves as vice president of sustainability and environment, health and safety (EH&S) for the Dow Chemical Company. In this global role, he leads Dow's sustainability strategy and team and drives implementation of Dow's transformational 2015 sustainability goals. Dr. Hawkins also leads Dow's global organizations for product safety, regulatory affairs, health services, EH&S auditing, and remediation. Dr. Hawkins joined Dow in 1988 and has served in a wide range of EH&S operations, and public policy roles across the company. Dr. Hawkins is also a recognized expert in sustainability business practices and environmental policy. He chairs the Strategic Advisory Council for the University of Michigan Erb Institute for Global Sustainable Enterprise and serves on the boards of Keystone Center, Global Water Challenge, World Environment Center, and Corporate EcoForum. He is also a member of the National Academies Roundtable for Science and Technology for Sustainability. He holds master's and doctoral degrees from Harvard University, School of Public Health, and a bachelor's degree from Georgia Tech.

Michael Kavanaugh, Ph.D., is a principal with Geosyntec Consultants, Inc., an engineering and consulting firm with offices throughout the United States and abroad. His research interests have included hazardous waste management, soil and groundwater remediation, process engineering, industrial waste treatment, technology evaluations, strategic environmental management, compliance and due diligence auditing, water quality, water and wastewater treatment, and water reuse. He has served as chair to the NRC Board on Radioactive Waste Management and the Water Science and Technology Board. Dr. Kavanaugh is a registered professional engineer in California and Michigan, a board-certified environmental engineer in water quality and sustainability for the American Academy of Environmental Engineers. He is an elected member of the National Academy of Engineering. He received his B.S. in chemical engineering from Stanford University, an M.S. in chemical engineering and a Ph.D. in civil and environmental engineering from the University of California at Berkeley.

Stephen Polasky, Ph.D., is the Fesler-Lampert Professor of Ecological and Environmental Economics at University of Minnesota. He received a Ph.D. in economics at the University of Michigan in 1986. He previously held faculty positions in the Department of Agricultural and Resource Economics at Oregon State University (1993-1999) and the Department of Economics at Boston College (1986-1993). Dr. Polasky was the senior staff economist for environment and resources for the President's Council of Economic Advisers 1998-1999. He was elected into the National Academy of Sciences in 2010. He was elected as a fellow of the American Academy of Arts and Sciences in 2009 and a fellow of the American Association for the Advancement of Science in 2007. His research interests include ecosystem services, natural capital, biodiversity conservation, endangered species policy, integrating ecologic and economic analysis, renewable energy, environmental regulation, and common property resources. He has served as coeditor and associate editor for the *Journal of Environmental Economics and Management,* as associate editor for *International Journal of Business and Economics,* and is currently serving as an associate editor for *Conservation Letters, Ecology and Society and Ecology Letters.*

Kenneth G. Ruffing, Ph.D., is an independent consultant and author specializing in sustainable development, environmental economics, and development economics. Among other consultancy assignments, he has advised the United Nations (UN) Environment Programme on the Green Economy Project, advised the Organisation for Economic Co-operation and Development (OECD) on sustainable development and on environmental aspects of policy coherence for development and served as coordinator of the African Economic Outlook from 2006 to 2009. He was formerly deputy director and chief economist of the OECD Environment Directorate from 2000 to 2005 where he took a special interest in the issue of decoupling environmental pressure from economic growth. Prior to joining OECD, he had a long career with the UN, beginning in 1971 while completing his Ph.D. dissertation, entitled *The effects of inflation on the structure and yield of the fiscal system of Chile,* at Columbia University. Dr. Ruffing has worked as a development economist for the UN Economic Commission for Latin America and the Caribbean; provided technical assistance in external debt management to developing countries for the UN Conference on Trade and Development; represented the UN at numerous debt rescheduling exercises carried out by the "Paris Club" of official creditors; was secretary to the UN Committee for Development Policy, where he provided technical expert support for 10 years; prepared the UN macroeconomic forecasts for the world economy based on Project LINK from 1989 to 1993; and served as deputy director for the UN Division for Sustainable Development for 7 years. During his long career with international organizations Dr. Ruffing has conducted research, undertaken scholarly reviews, and published articles on a wide range of sustainable-development and economic-development issues; environmental and economic development policy integration; natural resource economics (oil and water); macroeconomics, external debt and finance; trade, aid and development; development planning and its integration with public-sector budgeting; monetary and fiscal aspects of public policy in developing countries; and economic reform processes and economic convergence.

Armistead Russell, Ph.D., is the Georgia Power Distinguished Professor and Coordinator of Environmental Engineering at the Georgia Institute of Technology. Professor Russell arrived at Georgia Tech in 1996, from Carnegie Mellon University, and has expertise in air-quality engineering, with particular emphasis in air-quality modeling, air-quality monitoring and analysis. He earned his M.S. and Ph.D. degrees in mechanical engineering at the California Institute of Technology in 1980 and 1985, conducting his research at the Caltech Environmental Quality Laboratory. His B.S. is from Washington State University (1979). Dr. Russell has been a member of a number of the NRC committees, including chair of the Committee to Review EPA's Mobile Source Emissions Factor Model and chair of the Committee on Carbon Monoxide Episodes in Meteorological and Topographical Problem Areas. He also served on the Committee on Tropospheric Ozone Formation and Measurement, the Committee on Ozone Forming Potential of Reformulated Gasoline and the Committee on Risk Assessment of Hazardous Air Pollutants. Dr. Russell served on two EPA Science Advisory Board subcommittees: the Clean Air Science Advisory Committee's Subcommittee on the National Ambient Air Monitoring Strategy and the Subcommittee on Air Quality Modeling of the Advisory Council on Clean Air Compliance Analysis. He was also a member of the EPA Federal Advisory Committee Act Subcommittee for Ozone, Particulate Matter and Regional Haze and the North American Research Strategy for Tropospheric Ozone and California's Reactivity Science Advisory Committee. Previously, he was on the Oxygenated Fuels Program Review of the Office of Science and Technology Policy, various NRC program reviews, and a committee to review a Canadian NRC program.

Susanna Sutherland has a degree from the University of Tennessee in environmental studies with forestry minor, and an M.S. in biosystems engineering technology with an emphasis on water quality. She has worked with the Tennessee Department of Environment and Conservation and in the Tennessee State Park system. She also worked with the Tennessee Valley Authority, first in environmental policy and planning in both Alabama and Tennessee and later in river operations and environment. In 2007, Ms. Sutherland came to the city of Knoxville as the South Waterfront Development's project manager and, in 2009, became the city's program manager of sustainability. Her current responsibilities include implementing the city's U.S. Department of Energy grants, staffing the Energy and Sustainability Task Force, and chairing Knoxville's electric vehicle advisory board. Ms. Sutherland's overarching goals include incorporating efficient and sustainable best practices into municipal operations and promoting environmental responsibility in the Knoxville community as an economic driver.

Lauren Zeise, Ph.D., is Chief of the Reproductive and Cancer Hazard Assessment Branch of the California Environmental Protection Agency. She oversees or is otherwise involved in a variety of California's risk assessment activities, including cancer and reproductive toxicant assessments; development of frameworks and methodologies for assessing cumulative impact, nanotechnology, green chemistry and safer alternatives, and susceptible populations; the California Environmental Contaminant Biomonitoring Program; and health risk characterizations for environmental media, food, fuels, and consumer products. Dr. Zeise's research focuses on human interindividual variability, dose response, uncertainty, and risk. She was the 2008 recipient of the Society of Risk Analysis's Outstanding Practitioners Award and is a national associate of the NRC. She has served on various advisory boards and committees of the EPA, Office of Technology Assessment, the World Health Organization, and the National Institute of Environmental Health Sciences. She has also served on numerous NRC and Institute of Medicine committees and boards, including the committees that produced *Toxicity Testing in the 21st Century: A Vision and Strategy; Science and Decisions: Advancing Risk Assessment;* and *Understanding Risk: Informing Decisions in a Democratic Society.* Dr. Zeise received her Ph.D. from Harvard University.

Appendix B

Statement of Task

The U.S. Environmental Protection Agency's Office of Research and Development (ORD) has been working to create programs and has been examining applications in a variety of areas to better incorporate sustainability. To further strengthen the analytic and scientific basis for sustainability as it applies to human health and environmental protection, an ad hoc committee under the Science and Technology for Sustainability Program will conduct a study and prepare a report that will answer the following questions.

- What should be the operational framework for sustainability for EPA?
- How can the EPA decision-making process rooted in the risk assessment/risk management (RA/RM) paradigm be integrated into this new Sustainability Framework?
- What scientific and analytical tools are needed to support the framework?
- What expertise is needed to support the framework?

Appendix C

Glossary

Biocentrism: The belief that all living things have intrinsic value.

Breakthrough objective: Commonly referred to in the business community, breakthrough objectives are goals that extend far beyond the current capabilities and experiences of an organization and require new strategies and approaches to ensure successful attainment of these goals. These objectives are generally designed to improve performance throughout an organization.

Change management: A process to prioritize allocation of resources and to provide a mechanism for prioritizing any change so that critical changes are made first, followed by low priority changes. The process also ensures changes are implemented on a set schedule and is a collaborative process that requires representative involvement from as many stakeholder groups as is feasible.

Culture: the integrated pattern of human behavior that includes thought, speech, action, and artifacts and depends upon the human capacity for learning and transmitting knowledge to succeeding generations (Merriam-Webster 2001).

Goal: What is specifically sought to be achieved and is determined through the use of measured indicators.

Indicator: A summary measure that provides information on the state of, or change in, a system (OECD 2011b), i.e., what is being measured.

Integrated assessment tools: Tools that link in a consistent fashion formal models of the environment and society (NRC 1999).

Interdisciplinary: Approach that expands the multidisciplinary approach so that communication is more frequent and members are involved in problem-solving beyond the confines of their discipline (Dyer 2003).

Intergenerational equity: The fair distribution of costs and benefits among different generations.

Intragenerational equity: The fair distribution of costs and benefits among different groups of the same generation.

Knowledge management: Strategies that an organization uses to enable the creation of knowledge and to distribute this knowledge (OECD 2000).

Metrics: Defines the unit of measurement or how the indicator is being measured (OECD 2011a).

Multidisciplinary: Approach in which independent, discipline-specific members conduct separate assessment, planning, and provision of services within their own disciplines with little coordination of information (Dyer 2003).

Optimize: To select the best option from a set of possible alternatives.

Place-based: The use of a geographically defined area to integrate or coordinate programs. Projects that are based in a specific locale with measurable outcomes (Barca 2009).

Process: A systematic series of actions designed with a goal as the endpoint.

Resilience: The ability of a system or a community to absorb shocks and still retain the same basic structure and functions (USGS 2011).

Screening: The use of a model or analytic method designed to select which problems or decisions should be subject to further analysis (EPA 2011).

Sustainability: To create and maintain conditions under which humans and nature can exist in productive harmony and that permit fulfilling social, economic, and other requirements of present and future generations (NEPA 1969; Executive Order 13514, 2009).

Sustainability analysis: The identification and analysis of key factors that are likely to have an impact, either positively or negatively, on delivering sustainable benefits (AusAID 2000).

Sustainability impact assessment: Impact assessment where all three dimensions of sustainable development are integrated into one assessment procedure and where the interdependence of dimensions is analyzed before decisions are made (Berger 2008).

Sustainability principles: Idea that sustainability must balance the needs of three components or pillars—social, environmental, and economic.

Sustainability science: An emerging field of research dealing with the interactions between natural and social systems that seeks to facilitate a transition toward sustainability (Clark 2007).

Sustainability technology: Technologies that prevent, remove, and control environmental risks to human health and ecology (EPA 2010).

Sustainable development: Development that meets the needs of the present without compromising the ability of future generations to meet their own needs (WCED 1987).

Sustainable innovation: Designing and implementing sustainable organizational processes and practices that generate social, environmental, and economic worth for all stakeholders involved (van Osch and Avital 2010).

Tool: Something regarded as necessary to the carrying out one's occupation or profession (Merriam-Webster 2001).

Transdisciplinary: Approach that, through all steps of the implementation of a product, involves the widest span of disciplines to bring different perspectives to the table (Anastas 2010).

REFERENCES

Anastas, P. 2010. ORD: The Path Forward. Memorandum to U.S. EPA Office of Research and Development, from Paul T. Anastas, Assistant Administrator. March 5, 2010 [online]. Available: http://yosemite.epa.gov/sab/sabproduct.nsf/796BB04146A5F14C852576F9004E5E69/$File/Anastas+Path+Forward+3-5-10.pdf [accessed May 3, 2011].

AusAID (Australian Agency for International Development). 2000. Promoting Practical Sustainability. Agenda Item 10. 33rd Meeting of DAC (Development Assistance Committee) Working Party on Aid Evaluation, November 22-23, 2000. Organisation for Economic Co-operation and Development [online]. Available: http://www.oecd.org/secure/pdfDocument/0,2834,en_21571361_34047972_31950220_1_1_1_1,00.pdf [accessed May 3, 2011].

Barca. F. 2009. Pursuing Equity through Place-Based Development Policies. Presentation at OECD/TDPC Symposium on Regional Policy, December 2, 2009, Paris [Online]. Available: http://www.oecd.org/dataoecd/41/37/44305783.pdf [accessed May 3, 2011].

Berger, G. 2008. Sustainability Impact Assessment: Definition, Approaches and Objectives. Presentation at OECD Workshop on Sustainability Assessment Methodologies, January 14-15, 2008, Amsterdam [online]. Available: http://www.oecd.org/dataoecd/16/52/39924538.pdf [accessed May 3, 2011].

Clark, W.C. 2007. Sustainability science: A room of its own. Proc.Natl Acad. Sci. U.S.A. 104(6):1737-1738.

Dyer, J.A. 2003. Multidisciplinary, interdisciplinary, and transdisciplinary educational models and nursing education. Nurs. Educ. Perspect. 24(4):186-188.

EPA (U.S. Environmental Protection Agency). 2010. Sustainable Technologies. Office of Research and Development, U.S. Environmental Protection Agency [online]. Available: http://www.epa.gov/nrmrl/std/ [accessed May 3, 2011].

EPA. 2011. Modeling Glossary. Council for Regulatory Environmental Modeling, Office of Science Advisor, U.S. Environmental Protection Agency [online]. Available: http://www.epa.gov/crem/glossary.html [accessed May 3, 2011].

Merriam-Webster. 2001. Merriam-Webster's Collegiate Dictionary, 10th Ed. Springfield, MA: Merriam-Webster.

NRC (National Research Council). 1999. Our Common Journey: A Transition toward Sustainability. Washington, DC: National Academy Press.

OECD (Organisation for Economic Co-operation and Development). 2000. Knowledge Management: The New Challenge for Firms and Organizations, September 21 and 22, 2000, Ottawa, Canada [online]. Available: http://www.oecd.org/dataoecd/10/13/2667415.pdf [accessed May 3, 2011].

OECD. 2011a. OECD Guidance on Developing Safety Performance Indicators: Communities. Organisation for Economic Co-operation and Development [online]. Available: http://www.oecdsafetyindicators.org/node/565 [accessed May 31, 2011].

OECD. 2011b. Sustainable Development Glossary. Organisation for Economic Co-operation and Development [online]. Available: http://www.oecd.org/glossary/0,3414,en_2649_37425_1970394_1_1_1_37425,00.html [accessed Apr. 19, 2011].

USGS (U.S. Geological Survey). 2011. USGS Science and Decisions Center. Information Sheet. Ver. 1, Feb. 28, 2011.

van Osch, W., and M. Avital. 2010. From Green IT to Sustainable Innovation. Paper 490 in AMCIS 2010 Proceedings. AIS Electronic Library [online]. Available: http://aisel.aisnet.org/amcis2010/490 [accessed May 3, 2011].

WCED (World Commission on Environment and Development). 1987. Report of the World Commission on Environment and Development: Our Common Future. United Nations Documents [online]. Available: http://www.un-documents.net/wced-ocf.htm [accessed May 3, 2011].

Appendix D

Sustainability in the OECD

For the Organization for Economic Co-operation and Development (OECD), which comprises the world's developed countries, the basis of sustainable development is the successful integration of social, environmental, and economic policy (OECD 2001). In that spirit, the OECD plays several important roles in creating and sharing ideas and information regarding sustainable development and analyzing environmental and sustainability trends.[1]

The OECD provides an interpretation of key concepts in the sustainable-development literature, including interpretations that are consistent with mainstream environmental economics (Ruffing 2010). In 2001, the OECD secretary general issued a major report on sustainable development (OECD 2001). Among other things, the report argued for mainstreaming the concept of sustainable development into standard economic discourse and into the normal practice of governmental policies. The report took a capital-based approach to sustainable development—distinguishing between anthropogenic-made capital, natural capital, human capital and social capital—arguing that sustainability requires that the sum of these different types of capital on a per capita basis not decline over time. The report also acknowledges that the degradation of capital that has no substitute would lead to an irreversible loss for future generations (Atkinson et al. 1997, Neumayer 1999) and thus would require the maintenance of critical stocks of natural capital at a safe minimum level, an approach known as "strong sustainability." Similarly, the *OECD Environmental Strategy* (OECD 2001), which was adopted by the OECD environment ministers in 2001, interpreted a key sustainability concept by articulating four principles for the environmental pillar of

---

[1] For more details see http://www.oecd.org/topic/0,3373,en_2649_37425_1_1_1_1_37425,00.html.

sustainability, namely, regeneration, substitutability, assimilation, and avoiding irreversibility.[2]

The OECD also publishes reports on various topics related to sustainability and fosters dialogue and discussion on sustainable development among member countries, thus providing an opportunity for sharing and learning. For example, the OECD has prepared reports on institutionalizing sustainable development (OECD 2007), good practices in the National Sustainable Development Strategies of OECD countries (OECD 2006), and guidance on preparing sustainability assessments (OECD 2010). The Organization has been working on a "green growth" strategy for consideration of OECD ministers; the strategy would maximize synergies between ensuring environmental integrity and improving economic efficiency. In 1998, the OECD also established a roundtable on sustainable development, where environment and development ministers engage in informal dialogue on the international policy agenda of sustainable development.[3] The OECD also publishes regular environmental performance reviews of member countries.

The OECD prepares a variety of reports on global sustainability conditions. In its 2008 report, the *OECD Environmental Outlook to 2030,* the OECD made clear that many of the conditions that led to Earth Summit in 1992 still pose serious threats (OECD 2008). The report projected environmental and economic trends from the present to 2030 and recognized progress in addressing air quality, water quality, forestry, and waste management in developed countries. It also described "climate change, biodiversity loss, water scarcity, and health impacts of pollution and hazardous chemicals" as especially serious problems. Without new policy actions on these issues, the OECD said that "within the next few decades we risk irreversibly altering the environmental basis for sustained economic prosperity." The report also identified a suite of "achievable and affordable" policies for addressing these issues.

---

[2] OECD (2001) defined these terms as follows:

Regeneration: "Renewable resources shall be used efficiently and their use shall not be permitted to exceed their long-term rates of natural regeneration."

Substitutability: "Non-renewable resources shall be used efficiently and their use limited to levels which can be offset by substitution by renewable resources or other forms of capital."

Assimilation: "Releases of hazardous or polluting substances to the environment shall not exceed its assimilative capacity; concentrations shall be kept below established critical levels necessary for the protection of human health and the environment."

Avoiding Irreversibility: "Irreversible adverse effects of human activities on ecosystems and on biogeochemical and hydrological cycles shall be avoided."

[3] For more details, see http://www.oecd.org/pages/0,3417,en_39315735_39312980_1_1_1_1_1,00.html.

REFERENCES

Atkinson, G., W.R. Dubourg, K. Hamilton, M. Munasinghe, D.W. Pearce, and C.E.F. Young. 1997. Measuring Sustainable Development: Macroeconomics and Environment. Cheltenham: E. Elgar.

Neumayer, E. 1999. Weak Versus Strong Sustainability: Exploring the Limits of Two Opposing Paradigms. Cheltenham: E. Elgar.

OECD (Organisation for Economic Co-operation and Development). 2001. OECD Environmental Strategy for the First Decade of the 21st Century. Organisation for Economic Co-operation and Development, May 21, 2001 [online]. Available: http://www.oecd.org/dataoecd/33/40/1863539.pdf [accessed Apr. 18, 2011].

OECD. 2006. Good Practices in the National Development Sustainable Development Strategies of OECD Countries. Paris: OECD [online]. Available: http://www.oecd.org/dataoecd/58/42/36655769.pdf [accessed May 4, 2011].

OECD. 2007. Institutionalising Sustainable Development. Paris: OECD.

OECD. 2008. OECD Environmental Outlook to 2030. Paris: OECD.

OECD. 2010. Guidance on Sustainability Impact Assessment. Paris: OECD.

Ruffing, K.G. 2010. The role of the Organization for Economic Cooperation and Development in environmental policy making. Rev. Environ. Econ. Policy 4(2):199-220.

Appendix E

Sustainability Indicators

This appendix provides additional information about the extensive work nationally and internationally to develop sustainability indicators.

INTERNATIONAL EFFORTS

Reviewing recent international work on sustainability indicators, it is evident that measuring sustainable development has been a subject of many studies ever since the publication of the World Commission on Environment and Development's *Our Common Future* (WCED 1987). A comprehensive assessment of sustainability indicators was recently undertaken by the Scientific Committee on Problems of the Environment (SCOPE) (Hak et al. 2007). This assessment was followed by a collaborative effort on the part of statisticians in the United Nations Economic Commission for Europe (UNECE), Eurostat, and the Organisation for Economic Co-operation and Development (OECD) published as *Measuring Sustainable Development* (UNECE 2009). The approach taken in this report is a so-called capital-based one that examines ways to measure stocks and flows of economic (market-based), natural, human, and social capital. This approach is thoroughly discussed, drawing heavily on the Handbook for compiling the *United Nations System for Environmental and Economic Accounts* (UN/EC/IMF/OECD/World Bank 2003), commonly referred to as the SEEA where sustainable development is defined as "development that ensures non-declining per capita national wealth by replacing or conserving the sources of that wealth" (p.4); that is, stocks of produced, human, social, and natural capital.

Although this approach implies substitutability among the different types of capital, the authors drew on the literature of strong and weak sustainability to argue that some categories of natural capital should be defined as critical and thus not be allowed to fall below a minimum level: (1) a reasonably stable and predictable climate; (2) air that is safe to breath; (3) high-quality water in sufficient quantities; and (4) intact natural landscapes suitable for supporting a diversity of plant and animal life (UN/EC/IMF/OECD/World Bank 2003). Taking into account the difficulty of developing monetary measures for all the concepts needed, the authors argue for the use of physical measures and other proxies in developing a practical indicator set.

The authors noted that in actual practice, most sustainable-development indicator sets were policy-based—that is, capable of changing over a sufficiently short period of time to attribute the changes at least in part to specific policy measures. Examples of these types of indicators are emissions of greenhouse gases on an annual basis, energy use per unit of gross domestic product (GDP), mortality due to selected key illnesses. They identify 27 indicators used in indicator sets by at least 10 countries (see Table E-1). The authors do not question the usefulness of the policy-based approach, but strongly advocate that it be complemented by outcome-based indicators using the capital stocks- and flow-approach (UN/EC/IMF/OECD/World Bank 2003). Examples of these types of indicators are (1) (stock) average annual concentrations of ground level ozone, (flow) smog-forming pollutant emissions per month; (2) (stock) health-adjusted life expectancy, (flow) annual changes in age-specific mortality and morbidity, (3) (stock) real per capita natural capital, (flow) real per capita net depletion of natural capital. For more examples, see Table E-2.

The authors also reported a comparison of national and international sustainable-development strategies where they found that 11 indicator themes were common to a large number of the strategies: management of natural resources, climate change and energy, sustainable consumption and production, public health, social inclusion, education, socioeconomic development, transportation, good governance, global dimension of sustainable development, research and development, and innovation (UN/EC/IMF/OECD/World Bank 2003). The authors of the report suggested that most of the strategies can be captured by the outcome-based indicators they propose but that indicators related to efficiency and equity might well be added to the set to capture other dimensions of policy.

While the work mentioned above was under way, the United Nations Commission for Sustainable Development (UNCSD) revised its set of indicators in *Indicators of Sustainable Development: Guidelines and Methodologies* (UNCSD 2007). Because many of the same statisticians contributed to both exercises, there is considerable overlap, at least among the policy-based indicators. The UNCSD set now contains 50 core indicators among a larger set of 96 sustainable-development indicators. Among the 27 indicators mentioned above, all but one (unemployment rate) are included in the broader UNCSD set, and 20 of them are included among the core set.

TABLE E-1 Policy-Based Sustainable Development Indicators.[a]

Rank	Broad Indicator	Number of Indicator Sets Where Used
1	Greenhouse gas emissions	22
2	Education attainment	19
3	GDP per capita	18
4	Collection and disposal of waste	18
5	Biodiversity	18
6	Official Development Assistance	17
7	Unemployment rate	16
8	Life expectancy (or healthy life years)	15
9	Share of energy from renewable sources	15
10	Risk of poverty	14
11	Air pollution	14
12	Energy use and intensity	14
13	Water quality	14
14	General government net debt	13
15	Research and Development expenditure	13
16	Organic farming	13
17	Area of protected land	13
18	Mortality due to selected key illnesses	12
19	Energy consumption	12
20	Employment rate	12
21	Emission of ozone precursors	11
22	Fishing stock within safe biologic limits	11
23	Use of fertilizers and pesticides	10
24	Freight transport by mode	10
25	Passenger transport by mode	10
26	Intensity of water use	10
27	Forest area and its utilization	10

[a] Based on indicators where 10 or more countries or institutions have adopted them.
SOURCE: Measuring Sustainable Development, by UNECE, copyright (2009) United Nations. Reprinted with permission of the United Nations.

A large number of the capital stock, capital flow, and policy indicators discussed in these two recent reports can be found in OECD collections of data and indicators. In the various reviews and monitoring mechanisms for which such indicators are relevant, the policy-oriented and output-oriented capital-flow types of indicators are well represented. However, the capital-stock indicators are not widely used in OECD work; nor are they often used in the national sustainable-development indicators of the OECD member countries.

A large number of sustainability indicators have been identified, a large fraction of which being most directly related to the environmental pillar. The large number of environmental indicators may be indicative of the complexities

TABLE E-2 Outcome-Oriented Sustainable-Development Indicators

Indicator Domain	Stock Indicators	Flow Indicators
Foundational well-being	Health-adjusted life expectancy	Changes in age-specific mortality and morbidity
	Percentage of population with post-secondary education	Enrollment in post-secondary education
	Temperature deviations from norms	Greenhouse gas emissions
	Ground-level ozone and fine-particulate concentrations	Smog-forming pollutant emissions
	Quality-adjusted water availability	Nutrient loadings to water bodies
	Fragmentation of natural habitats	Conversion of natural habitats to other uses
Economic well-being	Real per capita net foreign financial asset holdings	Real per capita investment in foreign financial assets
	Real per capita produced capital	Real per capita net investment in produced capital
	Real per capita human capital	Real per capita net investment in human capital
	Real per capita natural capital	Real per capita net depletion of natural capital
	Reserves of energy resources	Depletion of energy resources
	Reserves of mineral resources	Depletion of mineral resources
	Timber resource stocks	Depletion of timber resources
	Marine resource stocks	Depletion of marine resources

SOURCE: Measuring Sustainable Development, by UNECE, copyright (2009) United Nations. Reprinted with permission of the United Nations.

involved in characterizing the environment, or social and economic indicators have had more time to mature. Therefore, such that a more refined set has been identified and is viewed as providing sufficient information. The large number of available indicators has resulted in a trend to reduce the large number by grouping indicators in indicator sets or to develop integrated or aggregated indicators, potentially all the way down to a single index of overall sustainability. An appropriately constructed indicator set should adequately characterize the state and potential trend of the environment at the scale applied. For example, OECD has reported on a set of aggregated indicators in *Aggregated Environmental Indices: Review of Aggregation Methodologies in Use* (OECD 2002).

Another effort to assess international sustainability indicators includes the Yale Environmental Performance Index. This index was developed to quantitatively assess a country's national performance on a core set of 25 performance indicators tracked across ten policy categories, including environmental, public health, and ecosystem vitality. The index is used to gauge how closely countries meet environmental policy goals (Yale 2011).

DEVELOPING AND SELECTING THE APPROPRIATE SUSTAINABILITY INDICATORS

Although a large number of indicators may make communication more difficult and have less impact, information is potentially lost as the number of indicators is reduced, and the process likely to involve some level of arbitrariness, which can in turn lead to argument. For example, the Columbia/Yale Environmental Sustainability Index (ESI) began with 76 data sets grouped in five components to derive 21 indicators, which were then equally weighted to develop the ESI. The authors readily acknowledge the limitations of the ESI and understand that achieving full consensus on the appropriate weighting will be difficult. However, they also note its utility to gauge current environmental conditions and the likely future trajectory. Excessive narrowing of the range of indicators has the potential for negative unintended consequences, including masking potential trade-offs important to a decision.

Potential attractions to having a limited number of indices (potentially one) include communication and spurring change. The formulation of the index or indicator set, while open to debate, can also be used to promote beneficial change and to engage the public in the process. Public participation may include discussion of what indicator sets should be used. The index and indicator set would be integrative (synthetic), retaining the links between the facets of sustainability rather than being simply aggregate and would reflect all three pillars.

More detailed suites of indicators can be used for high-level analyses and discussions, such as identifying important trade-offs. For example the air-quality index (AQI) is a simple number that reflects multiple pollutant concentrations that is being effectively used to communicate something about the state of air locally. How to improve air effectively (and thus improve the local AQI) requires the consideration of a much larger set of indicators and the application of various tools to identify how the system will respond to policies and how indicators are linked. Thus, one set of indicators is linked to what is most directly actionable (emissions), another set is used to assess characteristics of the state of the system (pollutant concentrations) germane to the desired end point (healthy air), and the AQI is used to assist communication.

Data availability will be key to the development and use of sustainability indicators. Data availability could be accomplished by EPA collecting indicator data but not providing the score card. Local agencies should compile their own report cards based on guidance put forth by EPA. Data is the foundation of both indicators and the application and evaluation of tools. The type of data that should be collected will be determined as a product of goal setting and the resultant choice of indicators and tools. The potential availability of data will also inform the choice of goals, indicators, and tools.

Data that are likely to be used will probably be characterized by heterogeneity and range across multiple environmental measures. Collecting environmentally oriented data and maintaining them in a readily usable system, such as EPA's

Regional Vulnerability Assessment (ReVA) Program,[1] would introduce a systematic way to address the heterogeneity. Data that would be used for economic and social analyses would probably be obtained from the associated agencies.

Potential areas where environmental sustainability indicators may need to be introduced outside of typical environmental indicators are those characterizing health risks associated with environmental exposure and environmental justice. Further, fully capturing the benefits of adopting a sustainability approach will require adding indicators addressing economic and social considerations. Sustainability indicators currently do include health end points that may be impacted by environmental stressors. However, these indicators (e.g., hospital admissions, cancer incidence) usually describe endpoints that reflect many causes of which traditional environmental problems are a small fraction. These indicators may, in part, be drawn from World Health Organization (WHO) analyses (e.g., the Burden of Disease); prior EPA risk assessments; and work done by the U.S. Center for Disease Control (CDC). The advantages of such indicators are that they can be directly integrated within risk assessment and tools and expertise that already exist within and beyond the agency.

In the field of human health, there has been many advances in the modeling of infectious diseases using geographical information and similar advanced tracking tools, which can include the many indicators that are part of sustainability. The indicators are now being applied to environmental health by the integration of health, exposure, and hazard information (CDC 2010). Other social and economic parameters, such as housing stock and income, can be added, as is being done in the U.S. National Children's Health Study of 100,000 children to be followed from early pregnancy to adulthood (Scheidt et al. 2009). However, the challenges to develop integrated indicators for environmental health are greater than for those infectious disease because the infectious disease cause and outcome are much more clearly linked (e.g., cholera is caused by *Vibrio. Cholera,* tuberculosis is cause by *Mycobacterium Tuberculosis,* and avian flu can be identified and tracked in birds and humans). For example, it is not known which cases of asthma are caused by ozone inhalation and which cases of bladder cancer are caused by arsenic in drinking water, thereby complicating the development of information needed to clearly link human health effects and the environmental risks managed by EPA.

---

[1] The ReVA Program conducts research on various innovative approaches to evaluate and interpret large datasets and uses models to assess the current conditions and probable outcomes of environmental decisions. Working with various decision makers, such as regions or national program the data are used to understand the current conditions of an area of interest. ReVA is used to conduct research on stressors that may be influencing those conditions and to develop scenarios to project how stressors may look in the future (EPA 2009).

SUSTAINABILITY INDICATOR EFFORTS AT EPA

EPA, because of its mission, has focused more on the state of the environment than on economic and social considerations. In this respect, it has made considerable progress in recent years. In its 2008 *Report on the Environment* (ROE), the agency provided historical trends and analysis on 85 indicators related to the environment and human health (EPA 2008). The indicators chosen were based on a set of six well-defined criteria (see Box I-1 in the ROE) and were used in the five thematic chapters of the report: Air, Water, Land, Human Exposure and Health, and Ecological Conditions. Among the 85 indicators, both stock-and-flow indicators are represented, and many of them, particularly the flow indicators, are highly policy relevant. As might have been expected, they have been well chosen from an environmental and human health perspective. Some indicators were meant to be updated roughly every 2 years, and a number of them to were be updated each quarter. By the end of 2010, all 85 indicators had been updated. The report stated that administrative, socioeconomic, and efficiency indicators were not included. By and large, sustainable use of natural resources is not addressed either, with the exception of an indicator on freshwater utilization. Thus, the ROE cannot be used to understand the relationship between social or economic drivers and environmental pressures.

A sustainability approach would require EPA to use indicators that would include those additional considerations. Although identification of indicators involves other stakeholders, identifying economic and social indicators in particular can be done collaboratively with other agencies whose missions focus on those issues. Environmental justice is a growing concern, but review of the typical indicators used to characterize either social or environmental states suggest that there is a need to develop indicators specific to this concern.

Environmental sustainability indicators will be derived directly from observations or analysis of observations (e.g., as a result of modeling). Both types of indicators have uncertainties, and information about the level of uncertainty may be useful in Sustainability Assessment and Management approach and resulting decisions.

REFERENCES

CDC (Center for Disease Control and Prevention). 2010. National Environmental Public Health Tracking Network. Center for Disease Control and Prevention [online]. Available: http://ephtracking.cdc.gov/showHome.action [accessed May 5, 2011].

EPA (U.S. Environmental Protection Agency). 2008. Report on the Environment. U.S. Environmental Protection Agency, Washington, DC [online]. Available: http://www.epa.gov/roe/ [accessed May 5, 2011].

EPA. 2009. What is ReVA. Regional Vulnerability Assessment (ReVa) Program, U.S. Environmental Protection Agency [online]. Available: http://www.epa.gov/reva/about.html [accessed May 5, 2011].

Hak, T., B. Moldan, and A.L. Dahl, eds. 2007. Sustainability Indicators: A Scientific Assessment. Washington, DC: Island Press.

OECD (Organisation for Economic Co-operation and Development). 2002. Aggregated Environmental Indices: Review of Aggregation Methodologies in Use. ENV/EPOC/SE(2001)2/FINAL, Apr. 26, 2002. Organisation for Economic Co-operation and Development [online]. Available: http://www.oecd.org/officialdocuments/displaydocumentpdf/?cote=env/epoc/se(2001)2/final&doclanguage=en [accessed May 5, 2011].

Scheidt, P., M. Dellarco, and A. Dearry. 2009. A major milestone for the National Children's Study. Environ. Health Perspect. 117(1):A13.

UNCSD (United Nations Commission for Sustainable Development). 2007. Indicators of Sustainable Development: Guidelines and Methodologies, 3rd Ed. New York: United Nations [online]. Available: http://www.uneca.org/eca_programmes/sdd/events/Rio20/WorkshopSDIndicator/SustainableDevelopmentIndicators.pdf [accessed May 5, 2011].

UNECE (United Nations Economic Commission for Europe). 2009. Measuring Sustainable Development. ECE/CES/77. New York: United Nations [online]. Available: http://www.unece.org/stats/publications/Measuring_sustainable_development.pdf [accessed May 5, 2011].

UN/EC/IMF/OECD/World Bank (United Nations, European Commission, International Monetary Fund, Organisation for Economic Co-operation and Development and World Bank). 2003. Handbook of National Accounting – Integrated Environmental and Economic Accounting 2003. Final Draft. ST/ESA/STAT/SERF/61/Rev.1. United Nations, European Commission, International Monetary Fund, Organisation for Economic Co-operation and Development and World Bank [online]. Available: http://unstats.un.org/unsd/envAccounting/seea2003.pdf [accessed May 5, 2011].

WCED (United Nations World Commission on Environment and Development). 1987. Our Common Future. Oxford: Oxford University Press.

Yale. 2011. Environmental Performance Index. Yale University [online.] Available: http://www.yale.edu/epi/files/2008EPI_Text.pdf. [accessed Apr. 25, 2011].